THE BODY CORPORATE

AN OUTLINE OF OUR SOLAR FUTURE

Paul Rigby

ISBN: 978-0-6486070-0-7 eBook
ISBN: 978-0-6486070-1-4 Paperback

BEGINNING

...creating a new theory is not like destroying an old barn and erecting a skyscraper in its place. It is rather like climbing a mountain, gaining new and wider views, discovering unsuspected connections between our starting point and its rich environment. But the point from which we started out still exists and can be seen, although it appears smaller and forms a tiny part of our broad view gained by the mastery of the obstacles on our adventurous way up.

— Albert Einstein and Leopold Infeld,
'The Evolution of Physics', Cambridge, Cambridge University Press 1971, p.31

There is a principle which is a bar against all information, which is proof against all arguments and which cannot fail to keep a man in everlasting ignorance; that principle is contempt prior to investigation.

— Herbert Spencer, Nineteenth Century English philosopher

First people will deny a thing; then they will belittle it; then they will decide that it was known long ago.

— Alexander von Humboldt, Eighteenth and Nineteenth Century German philosopher, explorer and scientist

NOT LOST IN THE STARS

Not very long ago, there was little indication the universe held any planets beyond the Solar System. Earth is just one of those that scientists now believe are even more numerous than stars. But there's also a massive multitude of stars; our planetary home is dwarfed by them. So are we *lost* in the stars?

Most of us don't think too hard about our position in the starry heavens; there are other matters closer to home likely capturing our attention. And perhaps we have no wish to focus on the unpleasant events we have been told could arrive from that direction – maybe an asteroid, colliding with Earth, such as the one that struck this planet around 66 million years ago and ended the reign of the dinosaurs. And what about a burst of greatly heightened sunspot activity – that could severely impact our communication systems and power grids?

Taking a less pessimistic view, scientists have long looked for life elsewhere in the Solar System and beyond, yet despite their perseverance they've effectively drawn a blank. So are we indeed, Lost in the Stars? Could we be merely an accident occurring in an uncaring cosmos?

It's hardly necessary to point out how unlikely it is that the vast majority would even consider such a question. Serious speculation about our place in the universe is left to a particular variety of scientist. The bodies that inhabit the night sky are perhaps just baubles vaguely associated with snatches of song or the occasional pretty show of colour, unless perhaps we make a hobby out of astronomy. There are so many issues taking precedence, issues that are becoming more and more vexatious, more and more distributed over so many countries and communities.

If as a race we really believed that planets and stars had anything to do with our day-to-day lives, then possibly... but with a whole basket of different problems bearing down, for the vast majority there is precious little attention or interest left available.

Humanity has not fared quite so well as we might have expected in say, the 1970s and for a short while thereafter. We were sure we had science on our side and it would keep us safe. Medical technology was developing new antibiotics, new drugs and new electronic aids; computers and better communications were making economies more efficient. Global politics was as always unpredictable, yet appearing to stabilise with the eventual demise of the Soviet Union. Yes, there was trouble in places like Bosnia, Ethiopia and Western Sahara but these ethnic conflicts would surely settle down? Some people were talking about potential problems of global warming and agricultural pollution, but not a great deal of action was taken. (Climate models have actually suggested the safe boundary for climate change was surpassed in the late 1980s.)

The new millenium has proved rather more tricky than many had believed possible. Because it has a strong bearing on the whole direction of this thesis, we might take an overview of some salient features including a few of the more obvious examples of increasing global difficulties. We should also try to balance the discussion with an understanding of how these difficulties are actually a precursor to something far more optimistic. Something we could even call *revelatory.*

Global events are becoming in line with a prediction made in 1978 that '...we will see an environment increasingly depleted of the resources to sustain life. You will see species after species being exterminated and becoming extinct because this is the function of the withdrawal force.' Withdrawal, described as one of the two universal polarities, with its complementary opposite being proliferation, was the driving force for these planetary disturbances. And further: 'The oncoming period is going to be more dreadful than you imagine.

You think chaos, surely; you think trouble, surely... but this oncoming travel is going to be more dire than is presently realised. '

There is already plenty of distress and misery caused by current chaotic change. Yet the upheavals can be viewed as essentially natural and even orderly: They have purpose. The purpose is renewal, the creature shedding its old skin to reveal a shiny new one, the chrysalis opening for birth of the radiant butterfly. We have heard these predictions before and they have not fulfilled their promise, but this one is backed by the same power that is causing our planet's current travail.

Because it's common for humans to focus on their immediate environment – family, finances, friends, food and shelter – we don't always notice what might be happening in the next continent or even around the corner. When there is bad news we can be quite adept at dismissing anything that doesn't affect us personally - call it a basic variety, or lower order, of psychological survival mechanism. But eventually, current events will catch up with even those who won't extend their antennae beyond the backyard.

While it is highly unlikely there has ever been a time when there were no global disruptions of any kind, the extent and depth of many current problems are becoming more severe. So let's consider a very few facts to remind us of the ominous state of the world situation, even though we may be reluctant to dwell on it.

The intention here is not fear-mongering, but simply to underscore the reality of our times and then to offer some more optimistic ideas and predictions.

Despite the increasing development of global connection and interdependence, mankind is currently at war. While one nation attacks its neighbour, another nation is at war with itself and mistreats, imprisons or kills many of its own citizens having a minority ethnicity, religious belief or political persuasion that doesn't suit the particular government. Political upheavals, frequently resulting in oppressive juntas, together with some savage civil wars, fill in this unpleasant picture.

The problems of massively displaced populations, famine and pandemics appear largely intractable. As at the middle of 2022, the United Nations confirmed there were some 100 million displaced persons, which computed to around one 78th to one 80th of the entire population of the planet. And very unlikely as it would have been not so long ago, even relatively wealthy but disillusioned Chinese citizens have been willing to risk negotiating the extraordinarily dangerous trail from Columbia to Panama, just attempting to enter the USA.

Even in more favoured countries, national leaders all too frequently fail to make the right calls for the wellbeing of their populations and the divide deepens between wealth and poverty. Some of us can still lead reasonably 'normal' and even luxurious lives, yet hundreds of millions cannot and the future prospects are not encouraging. (In early 2023 Oxfam reported that just under 700 million people were having to live on less than US$2 per day – hardly more than AU$20 per week.) Humanity's mental as well as physical health is – unsurprisingly – under increased pressure.

In mid-2022 the World Health Organization released its largest review of world mental health since the turn of the century. The WHO reported that there had been a 13% rise in mental health conditions and substance use disorders in the decade to 2017. 'In 2019, nearly a billion people – including 14% of the world's adolescents – were living with a mental disorder. Suicide accounted for more than one in 100 deaths and 58% of suicides occurred before age 50. Mental disorders are the leading cause of disability, causing one in six years lived with disability.'

And what of human health in general? While average global life expectancy more than doubled between 1800 and 2017 – from 30 to 73 years – the proportion of people's lives lived in poor or moderate health has remained unchanged at 50%. Medical science has made giant strides but there is no shortage of challenges; as we start to overcome one problem, another seems to pop up. A single aspect of this is the subject of a paper published in India by Shuvankar Mukherjee in 2017. The paper's abstract offered the following:

> Over the past 30 years, at least 30 new infectious diseases have emerged to threaten the health of millions of people across the globe. The major challenge to combat these infections is that for many of them, there is no specific treatment or cure or vaccine. There is limited scope of preventing or controlling them.

Medicine is itself part of the problem, the obvious example being that antibiotics are becoming less efficacious as we have used them ever more freely.

The recent global COVID-19 outbreak convincingly demonstrated that a pandemic creates a raft of problems on many levels and health authorities have warned us to expect more of such diseases. Should we extend our focus beyond the human situation, we can see a general trend emerge: the planet faces dramatic levels of species extinction.

According to science, there have been five mass extinctions on planet Earth and we are now experiencing the sixth, which is developing faster than any of the others did. Between 1970 and 2023, the populations of vertebrate species have declined by an average of 68% and currently more than 35,000 species are considered to be threatened with extinction. These figures come from the International Union for Conservation of Nature (IUCN).

The current rate of overall species extinction is up to thousands of times (not a misprint) higher than the historical average; estimated extinction rate at January 2023 was one species every five minutes. Australia is widely rated as having the worst extinction rate of mammals in the world; one in three of our native species are at risk of extinction. On the other side of the world, in 2020 researchers estimated that the European insect population had decreased 50% in the previous 50 years, with consequent effects on the pollination of food crops.

Another accelerating situation is climate change. It is too easy to become desensitised to the continuing references to this subject nowadays and it's perhaps understandable that many of us simply try to ignore it. Unfortunately, nature takes no account of our denial and the facts speak for themselves.

In February 2023, Antarctic ice fell to its lowest recorded level – 1.79 million sq km. This surpassed the previous record by 136,000 sq km – an area double the size of Tasmania. Less ice means more ocean water. The UN warned that an increase in the pace at which sea levels were rising threatened 'a mass exodus of entire populations on a biblical scale', according to Secretary

General António Guterres. Some nations could cease to exist, drowned under the waves. Guterres said the climate crisis was causing sea levels to rise faster than for 3,000 years, bringing a 'torrent of trouble' to almost a billion people.

One example of rising sea levels is Fiji, where communities near the coast face various difficulties. Some villages – along with national emergency resources – are being swamped. There are higher rates of disease as temperatures rise; increasingly destructive storms as oceans get warmer and disruptions to agriculture as the intrusion of saltwater damages existing farmland.

According to The Guardian newspaper in mid-2023, the Arctic was then warming three times faster than the global average. Summer sea ice had declined by 30% in the past 30 years; 90% of 'old ice', which is classified as five years old or more, had gone. That ice used to act as the 'great heat shield' for the planet, reflecting back the sun's rays. Now the loss of ice is producing a vicious spiral of heating.

Australia is widely accepted by our citizens to be a land of 'fires and flooding rains', but such events, driven by increasing climatic extremes, are now becoming monsters. Eastern Australia suffered massive destructive floods in 2021 and 2022, while in the Black Summer of 2019-2020, 24 million hectares of land was burnt (the area of the entire United Kingdom). Some 3000 homes were destroyed, three billion animals killed or displaced, 33 people died directly and some 445 more from smoke inhalation.

These bushfires-become-wildfires delivered an average estimate of 900 million tonnes of carbon dioxide into the atmosphere, while a study published in the scientific journal Nature found that smoke from the fires alone temporarily depleted the planet's ozone layer by 3% to 5%. Outside of more urban areas, fighting these types of fire events in Australia is to a great extent left in the hands of volunteers who temporarily give up their day jobs – and their sleep – to assist communities close or distant. (The NSW Rural Fire Service is the world's largest volunteer firefighting organisation.) A valuable and inspiring system, but it's in danger of becoming overwhelmed.

Unstoppable wildfires are almost as 'normal' now as they are devastating: California; Canada; Canary Islands; Croatia; Greece; Italy; Tunisia; Algeria; the Hawaiian island of Maui.

We can expect the many climate difficulties to worsen. We are powerless to stop 'acts of God' like earthquakes, tsunamis, cyclones or hurricanes, blizzards and tornadoes and their size and ferocity don't seem to be diminishing. Every time I've considered adding a new example of global destruction to this account, yet another and more current one hits the news.

Instead, let us recall an event from the recent past we might prefer to forget: the Japanese earthquake and tsunami of 2011 shook north eastern Japan. But the effects of the earthquake, the strongest in Japan's recorded history, were felt around the world, from Norway's fjords to Antarctica's ice sheet. Well over a million buildings were fully or partially destroyed, according to the Japanese Reconstruction Agency. Nearly 20,000 people were killed. The final total damage bill could reach $US235 billion.

Making this awful situation much worse, there was a cooling system failure at the Fukushima Daiichi Nuclear Power Plant, resulting in a level-7 nuclear meltdown and release of radioactive materials. One major difficulty was how to dispose of 1.3 million tons of water from the plant that had been contaminated. Starting in 2023 this water, which had undergone a custom purification system, was to be gradually released into the Pacific Ocean. The system works to extract radionuclides picked up by water as it had cooled the damaged reactors and fuel debris, as well as from rainwater and groundwater that got into the reactor buildings.

The Japanese earthquake exemplified the dual nature of our love affair with technology; one clever idea can create difficulties – unintended consequences – that we may then attempt to undo by devising another technological solution.

Another particularly modern phenomenon is the developing differences in human sexuality. This can be confusing and distressing to those who believe all aspects of the human form, having evolved over very many thousands of years, are complete and should never alter. We will find that this is quite wrong. In coming centuries, significant changes will continue to take place until the human form is no longer vertebrate or even 'solid'. We have much to learn about evolution.

As sexuality becomes less distinct and gender less stable, there are attendant problems including reactionary repugnance, blame and guilt from those who feel threatened by such changes. And over the past 50 years, human sperm counts appear to have fallen by more than 50% around the globe. If these findings are confirmed and the decline continues, it could have important implications for human reproduction. Of course, some people might regard such a development as a good thing, considering that Earth is now having to cope with eight billion humans for the first time.

There seems to be a peculiar disconnect as we consider the galloping increase in global population during the last few decades (up 266% from a 'mere' three billion in 1960) compared with the increasing possibility of the human race being decimated in the not too distant future due to the ongoing multiple threats. Why is this great population increase happening now of all times, when the planet is in many ways least able to sustain it?

This is a good question. Answering it necessitates acceptance of a number of premises: that non-physical aspects of mankind are aware of what we might call age changeover; humans wittingly or unwittingly assist the maturity of those aspects as they experience the ups and downs of life; certain developmental steps need to be completed for a new cycle, a new way of living, to unfold as required. This all adds up to pulling the trigger for human populations to flourish as they are now – just before the current world-wide situation ends in even far greater chaos.

Bigger populations certainly do not ensure greater resilience – and neither do those that are more tech-savvy. A recent study showed the addiction to smartphones has led to increasing numbers of accidents; people literally walk into walls, lamp-posts and even moving vehicles because 'screen time' is apparently more important than survival. And then a learned doctor claims that the fault lies squarely with the phone manufacturers...

The amazing development in communications during the last few decades has also unfortunately yet convincingly demonstrated a deep attraction to divisive, selfish motives as much as to our capacity for creativity and co-operation. The term '*social* media' is so often a heartless misnomer; with great swathes of misinformation that are proving to be decidedly *anti*-social.

For those who can see it (I'm sure there are some) the energies swirling in our planet must appear like something of a vortex, as proliferation and withdrawal vie for dominance.

Even money, one of the oldest pillars of human societies, is due to become a stickier problem. As we know, it can easily become an all-encompassing obsession of rich and poor alike. Monetary value is entirely dependent upon its acceptance as a medium of exchange, which will only work as long as there are goods and services both desired and offered. Should we reach a situation due to catastrophic circumstances where very little of what we need is actually available, then money is at risk of losing its age-long status. This scenario, though beyond the imagination of most people, certainly justifies the earlier quoted prediction of 'dire'.

While living conditions deteriorate, we see some contrary indications. There's benefit from those strong advances in helpful technology, as well as from continuing attempts by institutions and ordinary citizens to assist beleaguered communities at home and overseas in various ways. This includes helping refugees and the disadvantaged, as well as working to reduce pollution and take care of the environment. Much less obviously, there is an upsurge of those acting to develop a deeper spiritual understanding and practice; as a result finding inner resources as an antidote to depression and despair which helps others as much as themselves.

We should also acknowledge some heroic feats of national and international co-operation in dealing with the current crises. At other times, disasters only receive band-aids. Or else the assistance is simply not there.

While various attempts are made to investigate, identify and grapple with the snowballing and deadly global events, there is another factor that might not be spoken of but is an essential part of the changes underway.

New knowledge and understanding of ourselves – what is required of us and our destiny, unheralded by obvious means – is coming to humanity or at least, to some sections of it. Along with confusion and bewilderment there are fresh perspectives arriving. While this knowledge doesn't cancel out the tremendous problems on our doorstep, it does offer a different attitude and a pathway to a completely renewed way of life.

NO ACCIDENT

One of the oncoming perspectives is a very much deeper understanding of the relationship between mankind and the Solar System, a planetary grouping which is not simply our cosmic backyard but much, much more. The sooner we can come to a deeper appreciation of the intimate connection between humanity, Earth and the entire Solar System or Solar Body, the quicker we can gain the benefits of knowledge that is part of our bridge to the future.

At this point, the reader might well ask what possible connection could there be betwixt these two subjects – between matters literally of life and death, compared to something that sounds highly abstract, theoretical and literally 'not of this world'?

We will endeavour to bridge that gap by presenting a pattern of life that stretches all the way from the marvellously intricate human body to the outer reaches of the Solar System. The recommendation is to bid farewell to our conditioned, habitual thinking because there is a grand pattern within this system awaiting our understanding and acceptance.

And further, a precept underlying this document which should be stated now, is that there is nothing accidental or random in the universe. However difficult it might be to believe at first, the universe – or multiverse if you prefer – is highly ordered and unified rather than random and fragmented. And the assertion is not simply a statement of faith.

We can look at just two types of evidence in support of the hypothesis: Let's bring forward our first witness, John Stewart Bell. Despite hardly qualifying as a public figure, Bell is regarded as one of the most brilliant scientists to enter the field of Quantum Mechanics.

Born in 1928, he succumbed to a stroke in 1990, but had he lived another ten or twenty years Bell would doubtless have become far better recognised and very possibly, a Nobel Prize winner. Like many other scientists and intellectuals, Bell also had a quirky side. For instance, in 1988 he visited Belfast to lecture to the British Association. However, on recognising a man called Reggie Scott with whom he had worked as a technician over 40 years previously, Bell left the assembled dignitaries in order to have a yarn with Reggie about the old days when they were two young men in the teaching laboratory making their way in the world.

John Bell knew what he was a scientist *for*. While he accepted the value of laboratory experiments, he emphasised that limiting the work to laboratory operations '... is to betray the great enterprise. A serious formulation will not exclude the big world outside the laboratory.'

Bell is best known for his theorem, published in 1964, which proposed that either the statistical predictions of Quantum Mechanics or the principle of local causes is false. Another physicist who also worked in this area, Henry Stapp, declared in 1975 that Bell's Theorem was the most profound discovery of science. The Wikipedia entry on Bell's Theorem says: 'To date, all Bell tests have supported the theory of quantum physics and not the hypothesis of local hidden variables.' (One way to express this is the phrase 'locality fails'.) Science writer Gary Zukav commented that as a result we can infer: '...there really may be no such thing as "separate parts" in our world... In that case, the idea that events are autonomous happenings is an illusion.'

The implications are immense. In short, Bell's work strongly suggests that the universe is one stupendous entity – of an intelligence, capability and significance we can acknowledge in principle, while hardly able to grasp in its fullness. In other words, a living universe.

Now let's turn to a very different type of evidence for universal order: the work of Mark Vidler, who made observations of stars relating to terrestrial locations in the most unlikely way and presented them in his book The Star Mirror. I checked out his geographical placements (latitude and longitude) for most of his on-the-ground locations, although not his star positions. A divergence between us was that Vidler used Operational Navigation Charts for land areas – small-scale aeronautical charts that provide topographic information for air navigation; while I relied on the internet.

Here is what Vidler came up with and published in 1998: he discovered there were a number of very precise alignments between the positions of stars in the sky above planet Earth on the one hand and between features of the Great Pyramid at Giza, the positions of terrestrial mountain summits and ancient sacred sites on the other. We are faced with something that simply cannot be explained by what is usually considered 'logical' means. But there it is, available for anyone to investigate and confirm without needing any very special knowledge or training.

Throughout the Star Mirror book, there were a few unfortunate discrepancies and errors in what is likely an unfamiliar field of knowledge for many readers. Also, confusion could be caused by the lack of consistency in the given star names. This problem might easily have been laid to rest by explaining that there are frequently two names for the same star. One is its 'personal' name, for example Kochab, with the second name coming from its containing constellation and its luminosity, in this case Beta Ursae Minoris. (Or in English, 'the beta star of Ursa Minor' – 'beta' meaning it is the second most brilliant star in the constellation.) Furthermore, that particular constellation has two common names, the Little Dipper and the Little Bear, which if it was explained would also make things easier for readers. Despite these irritations, I found Vidler's research considerable and compelling.

Inspired by the initial work of construction engineer Robert Bauval, Vidler began his investigation by plotting the straight lines incorporated in the building of Egypt's Great Pyramid of Giza and the unlikely but very accurate relationships these lines had with significant stars in 2450BC. This is the date generally considered to be the year of the pyramid's completion, although that is not the only estimate of age: Giza still holds many secrets.

Intensive research by the former financial journalist Graham Hancock and his colleague Bauval, suggests a possibility that the Giza pyramid complex was built in more than one stage and that there could have been a period of thousands of years between the start and end dates. Yet interestingly, Hancock and Bauval nominating 2500BC as the most likely completion date, correlates rather closely with Vidler's discovery of multiple precise alignments between major stars and the Great Pyramid in 2450BC.

It is unrealistic to suggest that such alignments relied on happenstance. The relationships between most visible stars and the Earth's surface are constantly if very slowly on the move, due largely to the effect of precession, so 'luck' is hardly a viable explanation for the multiple results encountered by Vidler when using a computer program to hone in on this particular Egyptian era.

Precession is a word used to describe how planet Earth spins on its axis while that axis wobbles in the opposite direction or 'precesses' – just as does a plastic spinning-top that you could find in a Christmas cracker. The result is that visible constellations of distant stars tend to change their positions as centuries pass – a single degree approximately every 72 years or around 25,800 years to complete one circuit. (Polaris – the North Star – precesses more slowly because it is currently to be seen near Earth's axial centre. Other stars that are farther from the pole will appear to move faster in the sky due to precession.)

Following his initial research to confirm Bauval's original findings, Vidler then checked out star alignments to the pyramid from around the turn of the 2000 millenium. These alignments, although involving different stars from 2500BC due to the passage of precessional time, were very close. Is it too indulgent or too whimsical to ask whether something in the sky was celebrating an important upcoming event for planet Earth?

Next, Mark discovered more examples of precise alignments occurring in this modern era, between the brightest stars and major terrestrial mountains, as well as with ancient sacred sites – like Stonehenge, Tiahuanaco, Nazca, Easter Island and Glastonbury Tor. Almost impossibly, he found that many of the aligned stars were further aligned, within a close vicinity of sky, to additional stars that were related to the originals by forming the very unusual shape in the night sky of accurate isoceles triangles. (An isosceles triangle is one with at least two sides of equal length.) In addition he found a number of prominent, nearby topographical features that were also mapping out triangles of undeniably accurate isoceles dimensions.

Then, completely over-stretching the bounds of chance, Vidler started to find groups of three high points on the operational navigation charts that he used which apparently had nothing to do with star alignments, yet that once again showed precise isoceles triangle patterns.

Although I had to accept that these results were likely of little interest to people weighed down by day-to-day concerns, to me they were a light-bulb moment. The lesson was clear: there can be no coincidence, no accident in this universe. Everything is a totality of order.

Vidler wrote from his background as a semiologist. (Semiosis is any activity, conduct, or process that involves signs, where a sign is defined as anything that communicates something, usually called a meaning, to the sign's interpreter.) His intuition and enquiring mind came together to unwrap knowledge of what might otherwise, because hidden in plain sight, have been completely overlooked. He does an excellent job, painting a picture of his research and his mounting surprise and enthusiasm towards the findings. There is much more detail to the story than the bare facts presented here – interested readers can follow up for themselves and enter an alternate but self-consistent reality filled with wonder. Along with his discoveries, Mark's book shows how differently our modern society depicts the night sky, compared with how the ancient world understood it.

Graham Hancock had commented, in questioning the real function of the Great Pyramid, something that this author can readily confirm: '...it was also completely devoid of decorative features and of anything (figures of deities, reliefs of liturgical texts, and so on) which might be suggestive of worship or religion'. The bare, austere environment that clearly calls for a sense of reverence, is in complete contrast with the Egyptian Valleys of the Kings and Queens, which despite being cemeteries appear celebratory in their function of housing many finely crafted and colourfully bedecked tombs of the pharaonic ruling caste.

Vidler concluded that the Egyptian builders would have known exactly what they were doing, in completing the Great Pyramid to the incredibly precise dimensions, angles and alignments that they did. This is despite the common convention that Egypt in 2500BC had neither any knowledge of precession, nor the value of pi (approx. 3.14) which is to be found clearly incorporated in the geometry of the pyramid's construction.

Like Hancock, I was sure the building's function had nothing to do with a ruler's last resting place, which is only a convenient belief backed up by an unfortunate lack of credible evidence. Vidler had a slightly different approach:

'The Great Pyramid builders left nothing to account for themselves – no records, no plans, no formal communication at all. The Great Pyramid contains no hieroglyphic texts and no murals. It is apparently a mute monument. Yet here we see the pyramid pointing to stars, placing stars onto lines. Is this perhaps a form of writing, a manner of speech?'

In these observations we see an example of the connection with stars that has followed mankind's footsteps since time immemorial. But what of the planets?

PLANETARY PERSPECTIVES

The word 'planet' comes from the Greek word *planetes*, meaning 'wanderer', so called because Mercury, Venus, Mars, Jupiter, and Saturn – the five worlds visible to the naked eye – constantly and quite quickly change their positions in the sky compared to the stars, due to the planets' relative close proximity to Earth. The sun and moon were included in the term planet and the ancient Babylonians, Greeks, Persians, Romans and others thought of these seven Classical Planets as gods and named their seven days of the week after them.

Hertha Von Dechend asserted that planets had to be 'constitutive members of every mythical personnel'. Deschend co-authored Hamlet's Mill, described as 'An essay investigating the origins of human knowledge and its transmission through myth' – a work of no less than 505 pages in my copy. Georgio de Santillana, the other author of this old-styled, but meticulously-researched book, wrote that the Greeks 'called astronomy the Royal Science'. The authors argue that our myths are the remains of a pre-literate astronomy.

An extract: 'The most "ancient treasure" – in Aristotle's word – that was left to us by our predecessors of the High and Far Off Times was the idea that the gods are really stars, and that there are no others. The forces reside in the starry heavens, and all the stories, characters and adventures narrated by mythology concentrate on the active powers among the stars, who are the planets. A prodigious assignment it may seem for those few planets to account for all those stories and also to run the affairs of the whole universe.'

One might consider it compelling that the night sky held such special significance for many archaic civilisations, a significance that lingers even in our modern world in various forms of astrology. Like it, detest it or ignore it, astrology has been with us for many thousands of years, a fact which may give rise to some curiosity about the reason for its longevity.

Among the most creative theories to explain the human attraction to astrology, there is one that goes something like this: the planets have no direct influence in themselves, but are mirrors of basic organising principles in the universe. In other words, the fundamental patterns of the universe repeat themselves everywhere, in fractal-like fashion, giving substance to the familiar saying 'As above, so below'. (A fractal is a pattern that repeats forever. Every part of the fractal, regardless of how much one may be zoomed in on it, or zoomed out, looks very similar to the entire image.)

Sometimes, contemplating a completely unusual idea is a way to help in resetting the mind, leaving it more flexible for fresh input on a broader scale. We'll consider astrology more fully in later pages.

Similarly, one great benefit of learning more about the planets is that it tends to completely undermine our concept of 'normal', leaving us hopefully much more receptive to important new perspectives.

There are nine planets orbiting the sun – Pluto is included despite it having been downgraded by the International Astronomical Union in 2006 to the status of a 'dwarf planet'. There was not unanimous agreement with the IAU's ruling – at least two NASA senior executives, Jim Bridenstine and Alan Stern, added their voices to the dissenters. The justification for the downgrade was due to Pluto's inability to 'clear its neighbouring region of other objects'. Specifically, Pluto shares its orbital neighbourhood with Kuiper Belt objects such as the plutinos - objects that have orbital periods very similar to Pluto. It sounded like Pluto in its demotion was not to be taken seriously, instead labelled almost as cosmic detritus.

Despite this supposed humiliation, Pluto has plenty of engaging characteristics. It has five known moons -

Charon, Styx, Nix, Kerberos and Hydra. The largest of these is Charon, which is about half Pluto's size and one seventh its mass. Interestingly, Charon is tidally locked with Pluto such that the combination of the two is sometimes regarded as a binary object. And for a world that is around 170 times smaller than Earth, Pluto can still manage an atmosphere, even if it is mainly nitrogen with a little methane and carbon monoxide added.

We have to say at this point that 'divorcing' Pluto is untenable when we come to understand that the Solar System is neither a random collection, nor can it be a human selection of spherical bodies. To be explained.

Earth is the third planet out from the Sun and so placed as to enjoy what we regard as mostly a Goldilocks climate – neither too hot nor too cold. Our planet, a rocky world with the benefits of liquid water over much of the surface and gaseous oxygen in the atmosphere, is enough to make us shudder at the conditions existing on the others.

We might consider the almost unimaginable fluctuating extremes of temperature on Mercury, the furnace that is Venus and the crushing gravity, intense cold and poisonous atmosphere of Jupiter, the largest of the planets (and the one with the most moons – to date, 92 have been recorded). The other so-called gas giant, Saturn, would be equally inhospitable with its huge electrical storms extending over thousands of miles; lightning bolts on Saturn are 10,000 times stronger than those on Earth. Saturn's northern hemisphere has a raging, six-sided storm nicknamed 'the hexagon.' It's believed this towering, multilayered storm has been there for decades, if not hundreds of years.

The smaller ice giants, Uranus and Neptune, have other characteristics but are hardly more suitable for humans. Pluto, as we might expect due to its being furthest from the Sun at an average of 5.9 billion kilometers, is regarded as the coldest planet. (At its most severe, somewhere around minus 230 - 240 degrees Celsius.) And there is more to ponder over regarding this matter of planetary temperature.

Jupiter emits 67% more radiation than it receives from the Sun. We are told this is due to an internal heat source, thought to drive much of Jupiter's weather, which is another exercise in extremes. The planet experiences storms that can grow to thousands of kilometers in diameter within a few hours. The weather on Jupiter is so intense that it can actually be seen from space.

But what exactly is meant by the term 'internal heat source'? Here is one explanation:

Stars burn as a result of thermonuclear reactions deep in their cores. We are told that objects less massive than one percent of our Sun, like Jupiter, can never achieve the core temperatures required for thermonuclear reactions, let alone sufficient to develop into a full-blown star. Yet it could be said that Jupiter is, or has been, some way towards becoming a brown dwarf or what is called a 'failed star' which can generate an amount of energy from fusing deuterium into helium nuclei.

Incidentally, Jupiter's moon Io ('eye-oh') has hundreds of volcanoes and is considered the most active moon in the Solar System, sending plumes of sulphur up to 300 kilometers into its atmosphere. According to NASA, Io's volcanos emit, every second, more than 900 kilograms of gases and particles.

Returning to planetary temperatures, the ice giant Uranus radiates 1.1 times as much energy as it receives from the Sun; whereas Neptune radiates about 2.61 times as much solar energy as it receives.

Neptune is roughly 30 times as far from the Sun as Earth, so it gets correspondingly less heat and light. But it not only radiates far more heat than it takes in, this ice giant has much more activity in its atmosphere than planetary scientists would expect. It contrasts with the relatively nearby Uranus, which on average, depending on where it lies compared to Neptune's orbit, is around 1.6 billion kilometers closer to the Sun. Yet Uranus, despite receiving much more heat from the Sun, only radiates about the same amount of heat as Neptune.

The scientific community is baffled by the force of winds measured on Neptune, winds that reach 2,100 km/hour. The problem is, how can an icy cold planet like Neptune get its cloud tops to move so fast? One guess is that the low temperatures and the flow of fluid gasses in the planet's atmosphere might reduce friction to the point that it's easy to generate winds that move so quickly. Yet this is only a guess.

The closer we look at and into the planets, the more interesting or even bizarre they appear to be. For example: Uranus rotates on its side, with its ring system (yes, it has five rings, but not nearly as spectacular as Saturn's) positioned at an angle. The speculation is that this may have happened long ago due to a huge collision with another space object. Once again, we just don't know.

Uranus' unusual tilt causes what NASA considers to be the most extreme seasons in the Solar System. For about a quarter of each Uranus year (some 21 Earth years, as each Uranus year is 84 years long) the Sun shines directly over the north or the south pole of the planet. It means that for more than two decades on Earth, half of Uranus never sees the Sun at all.

These peculiar seasons are monitored and it was expected that the 2007 equinox on Uranus might cause unusual weather. But it was actually seven years later that the atmosphere erupted into wild, unpredicted storms, making Uranus more of a puzzle than ever. There are certainly plenty of opportunities for not only awe but also humility in the face of the many things we do not understand about the Solar System.

Closer to the Sun are the so-called rocky planets including Mars, Venus and Mercury.

Mars is little more than half the size of Earth but has some impressive topographical features. Gigantic volcanoes once dominated the surface, although none are active today. Olympus Mons is the biggest volcano to be discovered in the entire Solar System; at 602 kilometers across, its area is comparable to that of Arizona. At 25 kilometers high, this volcano is triple the height of Mount Everest. By volume, NASA tells us, Olympus Mons is 100 times larger than Earth's largest volcano, Mauna Loa in Hawaii.

No less than 4,000 kilometers long, the huge system of Martian canyons known as Valles Marineris is some five times the length of America's Grand Canyon. Valles Marineris was discovered by the global mapping mission Mariner 9 in 1971. The lack of active plate tectonics on Mars makes it difficult for scientists to understand how this giant canyon system was formed.

Venus is a planet that might reasonably serve as a model for hell. It has a devilishly high temperature and an atmosphere some 100 times thicker than the Earth's.

Supposedly, Venus is sufficiently close to the Sun that the little carbon dioxide it had in its early, Earth-like atmosphere caused the surface to warm up and leach out more carbon dioxide, creating a runaway greenhouse effect. Bone-dry and hot enough (around 472 Celsius) to melt lead, it's not exactly a welcoming environment for would-be astronauts. When the heavily shielded series of Venera spacecraft (only 10 survived a landing) from the Soviet Union touched down there in the 1970s and 80s, according to NASA each lasted a few minutes or, at most, a few hours before melting or being crushed beyond their ability to function.

Even above its surface, Venus has what we would consider a weird environment. Scientists have found that the planet's upper winds flow 50 times faster than the planet's rotation. The European Venus Express spacecraft, which orbited the planet between 2006 and 2014, tracked the winds over long periods and detected periodic variations. It also found that the hurricane-force winds appeared to be getting stronger over time.

Because it is so close to the Sun, we might well expect that Mercury is unimaginably hot. In terms of that planet's daytime temperatures, we'd be correct. During the day, temperatures on the surface can reach 430 degrees Celsius, yet because there is no atmosphere to retain that heat, night-time temperatures on the surface can drop to minus 180 degrees Celsius – more than a

600 degree difference within each very long Mercury day which is equivalent to 176 Earth days and just over two Mercury years! The axis of its rotation is tilted a bare two degrees with respect to the plane of its orbit around the Sun. Thus, Mercury spins nearly upright as does Venus, so they do not experience seasons in the same way as several other planets do. The greater the tilt in the axis, the more extreme the seasons.

(It's interesting to note that because the Earth's orbit around the Sun is not completely circular, in the southern hemisphere summer we're closer to the Sun than in southern hemisphere winter. So while the near-circular orbit has little effect on the Earth's climate, the seasons below the equator are actually more severe.)

Mercury is the second-smallest planet in the Solar System – after Pluto – and the second-densest after Earth. Mercury is possibly getting even smaller and denser. For many years, it was thought that Earth was the only tectonically active planet in the Solar System, but in 2016, data from the MESSENGER spacecraft revealed cliff-like landforms on Mercury known as fault scarps. (A fault scarp provides clues about the rate of tectonics and seismic quiescence or activity.) Because these fault scarps are relatively small, scientists believe that they weren't created very long ago and so they suggest that the planet is still contracting.

The Sun is at the heart of our Solar System – and we might well call that system a body. While the Sun's visible surface, known as the photosphere, is around 5,500 degrees Celsius, its upper atmosphere has temperatures estimated in the millions of degrees. Such a tremendous temperature differential comes with little explanation. All that I could find was a report, published in the journal Science in 2019, suggesting that geyser-like jets of magnetised plasma called spicules facilitate the transfer of energy through the sun's outer layers. Not much to go on.

There are other curious characteristics to this our local star: its equator, being composed of superheated gas or plasma, rotates nearly a third faster than its poles. Almost all known physical and biological cycles in the Earth's system are said to be driven by the electromagnetic radiation (Solar radiation) emitted by the Sun. Sunspots form when intense magnetic activity on the surface exposes the sun's cooler layers. Sometimes there are several sunspots – they can even occur in swarms – while at other times there are no spots. During a roughly 11-year period the number of sunspots goes from low to high and then back down to low.

When sunspots are active, more solar flares will result, creating an increase in geomagnetic storm activity for Earth. At these times, we will see an increase in the Northern and Southern Lights and a possible disruption to radio transmissions and power grids. In addition, I have to suggest the influence of sunspots will eventually prove to be of very much greater importance to all the nine planets than is currently recognised.

The Sun is far, far larger than any of the planets – it takes up 99.8% of the entire mass of our Solar System, with the remaining 0.2% being the planets, moons, asteroids, and other matter. By definition, the Solar System would not exist without the Sun. Organic life on Earth, as our scientists describe it, would never have appeared. No wonder primitive Man often worshipped this cauldron of heat, light and power, even if his prayers were left unanswered. But we shouldn't be too dismissive of this reverence, as there is much more to our local star than what we can discern as a result of applying our various technologies.

A simplistic model of the system might show the Sun as, say, a soccer ball and the planets as marbles circling it in orderly concentric circles; but the situation is more complicated than that. For instance, Pluto has an orbit around the Sun which is very elliptical; there are times when it crosses Neptune's orbit and comes closer to the Sun than Neptune does. From 1979 to 1999, Neptune was actually further from the Sun than Pluto, a phenomenon that happens every two and a half centuries. As Pluto's orbit takes it first closer to then further from the Sun, the frozen nitrogen on the surface alternately tends to sublimate and turn into a gas, or it re-freezes, which changes the atmospheric depth.

Although typically, the distance from one planet to another is taken to mean the distance between the two planets' orbits – and by that score Venus is Earth's closest neighbour – some scientists made an interesting discovery by simulating the planets moving in their orbits for thousands of years, while constantly calculating the distance between any two of them. They averaged those values together to find which planets are the closest to each other over time. The surprising result? Mercury was the closest planet to all the other planets (they excluded Pluto from their experiment). This might seem impossible, but we must take into account that every planet spends about half its time on the opposite side of the Sun. In Earth's case, while Venus does get very close to Earth it also spends plenty of time very, very far away.

We're told the closer a planet is to the Sun, the stronger the Sun's gravitational pull on it and consequently, the faster the planet moves. The farther it is from the Sun, the weaker the Sun's gravitational pull and the slower it travels in its orbit. Here, we introduce a new idea to some of the other specifics for each planet: its placement relative to the Sun including the degree of orbital ellipse, its revolution upon its axis, the angle of that axis of rotation to the plane of its orbit, as well as the speed of rotation. Repeating an earlier theme, they do not occur randomly or by accident; rather each planet contains living forces which require and obtain a precision in these matters.

In this brief review of the planets, we have looked at some pretty scary elements of our immediate cosmic neighbourhood. Yet while it is surely certain the other planets would not, could not, allow for what we define as living beings from Earth to survive unprotected in their particular environments, neither would these other planets' life forms be able to survive or flourish on our beloved Earth. No doubt this statement poses more than one challenge to our accepted current knowledge. We'll talk more about it later.

Despite all the unusual characteristics to be found in this Solar Body, there is clearly an overall balance holding the entire system within an acceptable framework and it has done so for an enormously long time. Such homeostasis is a feature of living systems, even if the idea of a living universe is still too ambitious for most of mankind. It's interesting to note that our planet's so-called primitive cultures tend not to impose that strict dividing line between our definition of organic life and features of the environment such as mountains, rock formations, waterways and seas. For some of these cultures, significance is everywhere and is to be honoured because at some level, they see that such significance *is* indeed life. As do many people on spiritual paths who recognise all energy, all matter, as alive.

Sir Isaac Newton (1643 – 1727) often described as the father of modern science, may not have agreed with that last sentence, but he *did* believe that the elegant and well-ordered Solar System must have arisen through both the divine *design* and the ongoing *dominion* of God. He hinted that this infinite dominion of the One God, in His omnipresence, can be linked to the universality of the phenomenon of gravity: Just as God is everywhere, so is gravity; the omnipresence of the One God upholds the one (*uni*versal) phenomenon of gravity.

At its best, science is an evolution and even fundamental concepts should never be ring-fenced. Recently, our scientific community saw an example of significant change with the announcement that the late, famous physicist Stephen Hawking in 2002 decided he had been wrong in his hugely best-selling book, A Brief History of Time. With his collaborator, the cosmologist Thomas Hertog, Hawking set about preparing a new way to encapsulate their ideas on the universe. Five years after Hawkings' death in 2018, Hertog explained: 'The problem for Hawking was his struggle to understand how the universe could have created conditions so perfectly hospitable to life.'

We might suggest that the solution to this 'problem'

has been staring the scientists – as well as the rest of us – directly in the face: That in fact, the universe is an expression of the One Life. Frequency is the key to this unity and the range of frequency has no known limit.

Hawking's final theory was eventually born and, in conjunction with Hertog's own analysis, they form the basis of On the Origin of Time, a book that takes its title from Charles Darwin's On the Origin of Species. Hertog said, 'It leads to a new philosophy of physics that rejects the idea that the universe is a machine governed by unconditional laws with a prior existence, and replaces it with a view of the universe as a kind of self-organising entity in which all sorts of emergent patterns appear, the most general of which we call the laws of physics.' He explained further, 'In the end, we both came to think of physics in a way much more like how we think of biology. We have put physics and biology on the same footing.'

From a press review of the new book: 'Examples of these life-supporting conditions include the delicate balance that exists between particle forces that allow chemistry and complex molecules to exist. In addition, the fact there are only three dimensions of space, permits stable Solar Systems to evolve and provide homes for living creatures. It is argued by some cosmologists that without these properties, the universe would probably not have produced life as we know it.'

(Albert Einstein is not only famous for discovering that energy and mass – matter – are interchangeable or in other words, different expressions of the same thing. He also articulated something not so very different from Hawking's difficulty: 'What is inconceivable about the universe, is that it should be at all conceivable.')

Science does have the benefit of *evolving* despite the more reactionary forces which don't want established ideas to be questioned. For example, Charles Darwin's work started a revolution in biology, and over the years some of his ideas on evolution of species have now coalesced into a widespread belief, or at least a vague acceptance, that the subject is 'nailed down'. The usual definition/description for this theory/belief today is Neo Darwinism.

But strangely, the more you look into it, the harder it is to find a single, clearly defined scientific explanation that is both accepted and corroborated across the board. You even begin to wonder about emperors parading with no clothes.

In 1964 the popular science writer Martin Gardner published The Ambidextrous Universe. The book was revised and re-published several times, including in 1967 and 1969. At least one version included the following statement: 'Evolution is simply the process by which chance (the random mutations in the nucleic acid helix caused by natural radiation) co-operates with natural law to create living forms better and better adapted to survive'.

Yet not all the common conclusions about the evolution of species are now tenable. They may not even have been tenable around the time Mr Gardner wrote the words quoted above. For example:

David Drayer, on the website Socratic.org, is described as having multiple credentials in the sciences and in teaching for over 40 years. He posted the following:

'Mutations are thought to be random changes in the information code of DNA that result in new information and life forms… Neo Darwinian evolution or descent with modification, postulates that starting with a relatively "simple" protocell, mutations caused the increase in diversity and complexity observed in the world.

'The Mathematical Challenges to Neo Darwinism, held in Philadelphia in 1966, (*a Symposium held at the Wistar Institute of Anatomy and Biology, April 25 and 26*) concluded that it is extremely difficult to assemble a new gene or protein by chance. For example, a relatively short protein of 150 amino acids has one chance out of 10 to the power 195 changes of occurring by random changes (mutations).

'The present empirical evidence in the study of DNA and the nearly impossible mathematical odds of creating new genetic information by random chance makes mutations an unlikely source of novel and more complex information as required by Neo Darwinian Theory. Mutations as a source of new information is vital to Neo Darwinian evolutionary theory. Present empirical evidence *does not support the necessary role of mutations in Neo Darwinian evolutionary theory.*' (My italics.)

No doubt this '10-with-195-noughts-chances-of-representing-a-single-chance-of-mutation' is sounding vastly less believable now as a driver of evolution. Instead, 'Survival of the Fittest' – a term coined by Herbert Spencer although later, Darwin also used it as well as his own description, 'Natural Selection' – appears to have been upgraded as an explanation for the huge range of planetary life-forms. It's a big ask for this process to have been responsible for the millions, billions or even a trillion species with all their beauty, their complex characteristics and their intricate functions that have proliferated across the globe over the last few hundred million years. (Consider the delicate, highly artistic forms of sea-horses. To describe these designs and the creatures themselves as a more or less happy accident is like believing that a team of monkeys with typewriters could eventually write all the works of Shakespeare.)

It's also fascinating to note that Sir David Attenborough, who has presented so many beautiful TV shows on our planet's fauna and flora, frequently explains to us that such and such a species has evolved in this or that direction, but never explains how it was done. Presumably we are left to assume that Neo Darwinian Theory is the only possible explanation. And how much easier is it to rest on that assumption, than produce a rigorous review of the evidence – which does not back the theory. After all, why complicate a much-loved TV series?

None of our neighbouring planets in the Solar System holds out much likelihood of having what scientists currently understand as life, evolved or not. (Yet some do retain hopes that Mars and Jupiter's three large, underground ocean-bearing moons – Ganymede, Callisto and Europa – might hit the jackpot). Mars, despite also showing evidence of sub-surface water, is still pretty inhospitable from our point of view. The planet has a barren surface, an atmosphere mostly of carbon dioxide and less than 0.2 percent oxygen, while average temperature is minus 63 degrees Celsius. Nevertheless, some astronomers will relentlessly continue to search for their holy grail. Yet they only want to find obviously organic life, as it is described and evidenced on our own blue planet.

In spite of the diversity of life forms on Earth, there is, nevertheless, a discernible patterning through these forms. But can we ever assume that there aren't other patterns, beyond Earth, that we are yet to discover and acknowledge?

This is something that has puzzled me (and hopefully

others) for years: Why do we generally insist that life means only the conventional life forms we see around us on this single world, in this tiny corner of the galaxy? I rate it an overly romantic attitude that fits all too snugly with the Lost in the Stars scenario. There is a rather obvious and more believable alternative – that because this *is* a living universe, Life will express Itself in a manner that suits each specific, individual energy environment, whether that manner be supposedly organic, inorganic or something else altogether. Additionally, its qualities and capacities will be tailored to the particular function it is designed to perform.

In passing, we could mention that Plato around 360 BC produced his book Timaeus and had his character of the same name speak the following words: '…using the language of probability, we may say that the world became a living creature truly endowed with soul and intelligence by the providence of God'. This demonstrates that a living universe is hardly a new proposal. There is even a name for this point of view – hylozoism – meaning no inorganic matter. In other words, it is the doctrine that all matter has life. But we could take this idea further: There is in fact *only* life and nothing else.

If we take that position on board, we might also acknowledge the fact that the various expressions of Life cannot be orphans. There will, there must, be connections. All is related, whether or not the relationships are easy to unlock. There can be communities and/or hierarchies of them, interacting as they must.

As a vital component part of this, we need to discuss cycling.

ETERNAL CYCLING

Cycling is a feature of the universal life; so closely integrated in our day-to-day activities that we can easily take it so much for granted that we overlook its existence, operation and importance.

The female menstrual cycle is obvious but even more so is the cycle of our planetary seasons, our day and night, our youth and old age, our living and dying. More recently there have been some indications of physiological cycling called biorhythms, although the theory is disputed by the mainstream. Cycles rule both recorded and assumed history, from the rise and fall of ancient and modern empires to much further back, where we are looking at the great ages of Earth's changing biology. We might here revisit the story of the fifth mass extinction – that in one particular period (the Cretaceous) there were many varieties of dinosaurs and in the next (the Paleocene) the dinosaurs were all gone, superseded by growing numbers and varieties of the upstart mammals.

So cycling is simple, isn't it? Not necessarily. Five major planetary cycles have passed since the beginning of

Earth's existence and although there is some degree of correlation between these cycles and the great biological eras – or more accurately, it is the other way around – the full significance of global cycles has gone almost entirely unrecognised. And it only could be recognised within a framework of clear vision and acceptance of the living universe, overseen, controlled and developed by non-physical (we might as well say spiritual) energies. If that sounds a little too vague, we could define those energies as comprising extremely high frequencies. For example, whereas we humans consider light as having a very high frequency, in universal terms, visible light has an exceedingly low frequency range. We are looking here at a future development and understanding of physics.

Cycling has pivotal status in the universe; we include all the planets orbiting the Sun and even the Sun itself. The action of planetary orbiting is an example of cycling in itself – the spatial placement being the result of energy causing it.

As Einstein indicated, the basis of everything is energy. He might well have put it something like this: 'Everything is energy and that's all there is to it. Match the frequency of the reality you want and you cannot help but get that reality. It can be no other way. This is not philosophy. This is physics.' We should add that how energy is caused to interact with itself, gives rise to change without surcease. Cycling is a principal feature of this constant alteration and transition.

It is proposed that a cycle – which also includes all sub-cycles within sub-cycles – always has a function; cycling and functioning fit together as a universal constant. Frequency is a key to this view. Frequency, that familiar up-and-down squiggle, can be seen in terms of polarity reversal. If we take the example of a musician plucking a string of a guitar, the action creates a sound wave which is a vibration and this movement could be described as energy moving in opposing directions. It could also be described as the dual requirements of that wave through which its function (a particular note) will be effected. However, something else is needed if the function (the required effect) of the process is to be attained – someone has to hear it. Thus we have duality and we have cycling in this intermingling of forces; that of the note firstly with that of the guitar and then with the human ear. But if the would-be listener is deaf, this is non-resonance which means both the cycle and the function is left unfulfilled.

There is more to tell about cycling. (First, a necessary definition: Our use of the term *Logos* here means a channel, a force of almost unimaginable intensity, which cannot begin to be understood consciously. A Planetary Logos is the top level in energy of any particular planet; we are restricting that term to planets of our own Solar Body.)

The cycle of all major functions such as performed by a Planetary Logos, will comprise multiple sub-cycles and in the unfolding process of the function, many of these sub-cycles will fail to develop as expected by the entity responsible for carrying out the function. What happens is that the function in its internal operation is impacted on by itself and as a result, the function travels a different path – nevertheless this is the path which it was always required to travel. Such an event is not a mistake but the subtle application of duality. Always there is the path which *could have been* travelled and the path that *is*. The consequence is distress in the affected entity, which humans would deem to be sickness but universal intelligence would not. (The reason for this disparity in description is that the higher powers see the situation in a more basic, mechanical aspect – it is a lack of energy balance.)

The effect of such distress in the planet will channel back into the Sun and have a blanketing effect, causing an ageing of the star. When the same sequence of events happens in human life, it will act as a major factor in the ageing of the incarnate body.

Cycles by definition have a beginning and end. Time is involved but not necessarily as we would expect. The timescale of each function is its own; we cannot expect any two functions to have the exact same experience of time because no two functions are precisely the same. There is one further consideration that is by no means minor – in strictly fundamental terms, time does not exist. The apparency of time is due only to function; function determines how time is going to present itself in that particular instance.

Returning to the great ages of Earth, our planet is afforded seven of them, although the final cycle is in fact the entire cycle and the other six are sub-cycles. We are currently in the changeover from cycle five to six, with the whole gamut of heightened attendant upheavals. However, we must not think we are picked out for poor treatment, because massively disruptive events have distinguished all the previous changeovers.

Further, the objective of the mighty cycles of Earth is never simply destructive. Their purpose is the introduction of different qualities that result in fresh development of the living planet. Each cycle has a powerful motif that turns a new page of planetary expression. Thanks to evolvement of the planet's energy signature, all is affected including the bringing forward of physical refinements to the environment and life forms. Yet to introduce this sweeping change, the existing structure cannot continue as before and consequently we experience the extensive turmoils that characterise age changeover.

The cycles together give rise to their own larger cycle which determines the complete experience of Earth as a functioning entity. The sequence is from purely non-physical energy to manifestations of greater and greater complexity until a certain point is reached that determines the apex of the cycle. Then there is a downswing where proliferation gives way to withdrawal; the point is eventually reached where materiality is utterly extinguished. While this might almost seem like an exercise in futility, the important point is that the process creates difference in the underlying energies, something which we might term the meaningful essence of the universe.

The concept of planetary cycles is by no means new. Historian R.T. Rundle Clark (1909 – 1970) who was a specialist in the religious ideas of ancient Egypt, observed, 'Underlying all Egyptian speculation is that time is composed of recurrent cycles which are divinely appointed'.

This idea may sound highly unusual to our current view of ourselves and our future; on the other hand it will be joined by many more as we progress further into the sixth cycle. The fresh interpretations will eventually be generally accepted. This is inevitable because all of

Earth's planetary systems will be adjusted to the terms of the current cycle and there is no possibility of anyone or anything being able to 'opt out' of their own accord.

Mankind will eventually relate to his changed living conditions and himself; and although the journey will be anything but easy, eventually we will find it worthy of gratitude. Outrageously utopian as it might seem to us now, there will be a future era inclusive of an atmosphere of 'sheer joy' when humanity is able to realise its true energy placement.

The great cycles of Earth were and are, not of the same duration. The first, which began before what we call organic life was established, was almost certainly the longest, measurable in hundreds of millions of years. The next four were of increasingly shorter duration, as the Logos of the planet became more mature in its ability to manage and develop the stupendous complexity of its planetary systems. Now we come to the sixth cycle, which is expected to last a mere 10 or 20 thousand years. And what happens once that time has expired? Then, we will have finished our run on the planet. Man will almost certainly become extinct in the terms that we now describe and experience extinction; yet the livingness of Man may eventually have a renewed and possibly even glorious path that only the future will divulge.

We mentioned patterning earlier and cycling is a pattern in itself. For example, in the relative microcosm of Earth there is human conception and there has also been its macrocosmic equivalent – the birth of the other planets within our Solar System. Whether for a baby or for a planet, the beginning of manifest life includes a period of conflict and confusion. The reason is that even when conception has successfully occurred, the new creation has yet to be stabilised.

This uneven beginning has to do with the twin forces we mentioned earlier; the forces that operate everywhere: on the one hand proliferation and on the other, withdrawal. The first force encourages expansion and ultimately includes physical existence from the underlying energy base, while the other acts to draw back from its manifested state or states and instead to reside in the fundamental potential that is the mysterious omnipresence we call energy.

Let's consider how most physicists currently think the universe, including our Solar System, came into being. The Big Bang theory is a cosmological model of the observable universe proposing that all matter came into existence at the same time, roughly 13.8 billion years ago. It states that at this time, matter was compacted into a very small ball with infinite density and intense heat called a Singularity, which then grew to form the still-expanding universe we have today. Stephen Hawking is credited with the idea that this Singularity was also 'infinitely small'. (Infinitely interesting, one might say.)

An alternative to the Big Bang was proposed by Albert Einstein in the 1920s. Einstein suggested an infinite series of oscillations, each beginning with a Big Bang and ending with a Big Crunch when manifestation ceases and rolls back into the originating energy. The Big Bang could then be seen as the proliferating half of the current cycle. Although it isn't very popular with scientists, Einstein's model (which even he was reported to have abandoned in the end) certainly fits with the same concept that has come from ancient sources such as Hindu and Buddhist traditions. These teachings include many references to vast time periods – such as kalpas, or days and nights of Brahma (each being 4.32 billion years). During the 'days' life materialises and in the 'nights' everything, including gods and men, 'sleeps'.

There might have been a Big Bang event but did this simply take place in a vast and growing void which, even now, may well continue to 'fill up' with more galaxies? And if so, how did that empty space get to exist? Or do we mean that space simply appears, even if there isn't anything to fill it? But what if space is not a convenient and ubiquitous 'gap' ready to be occupied?

A scientific forum was held at The American Museum of Natural History in Manhattan on March 20, 2013.

A group of physicists gathered to discuss the topic of *What is Nothing?* While the physicists did not reach a consensus on the main question, they did agree on one thing:

'The first, most basic idea of nothing – empty space with nothing in it – was quickly agreed not to be nothing. In our universe, even a dark, empty void of space, absent of all particles, is still something.'

We might then ask, what if space is an entity in its own right? What if space has to be created before any 'stuff' can be put into it? What if there is utterly nothing – and we do mean *nothing* – unless it is intentionally created by a great, Unknowable source?

Such speculations are more likely to engage us as the sixth cycle proceeds. This is in line with how physics has developed, from in-depth questions to some light appearing that illuminates possible answers, leading to further research and experimentation. Yet we should emphasise that none of this process ever happens by chance. Anything we truly learn will always be only what we are permitted to discover. And so the universal order is impeccably maintained.

So far, we have described the changeover from the fifth to the sixth great cycle mostly in terms of tribulation and cataclysmic change, but eventually there will be a positive result, including something we might only dream of: No more war on planet Earth!

Along with this momentous development, there will be the likely demise of great cities with all their wealth and poverty, their high culture, high pollution and high crime of various descriptions. Mankind will eventually be living a communal existence and have a far deeper understanding of universal themes. Our technology will no longer be appropriate or necessary to the enormously changed conditions we will encounter.

Another vital and necessary change to occur as the sixth cycle progresses, is acceptance of the close relationship between humanity, planet Earth and the Solar System. From being what might now be considered a theory on the far fringes of belief, it will become a welcomed fact.

BODIES MICRO AND MACRO

How many organs are to be found in the human body? (But why are we talking organs when we are supposedly discussing the Solar System? Firstly, we could reiterate that this system is also a body. We did say we intended to present a pattern. We did consider that pattern could throw much light on the situation of planet Earth at this point in its history. So we continue.)

A very quick internet check gave us various answers to our question of how many human organs there are; those answers ranged from numbers five to 78, which clearly has a lot to do with how one defines the word 'organ'. It appears we could, for example, call each of our teeth an organ. But here is the answer we would suggest – there are nine essential major organs, such that should any one of them fail, the body will die: they are the blood, kidneys, liver, lungs, digestive system, heart, marrow, skin and finally, flesh and muscles.

Although the brain and spinal cord might be included with much more justification than a single tooth, these two marvels of biological engineering do not in themselves ensure physical life. Rather, they facilitate the working of the organs which maintain the body's operation. The brain performs all the neurological functions that allow the body to maintain blood pressure, heart rate, body temperature, hormones and breathing. Yet the brain cannot do anything to repair terminally damaged organs.

There is a correspondence existing between the nine planets of our Solar Body and the nine organs of our physical body – one instance of the phenomenon we called patterning which exists right across the universe. Likewise, we note there is a strong pattern – broadly applicable but not exact – between the Sun and the human brain.

Humanity is an expression of the Solar Body so we should hardly be surprised that planets and our organs, as well as Sun and brain, display this correspondence.

The postulation is that the Solar System didn't just happen, the planets didn't coalesce willy-nilly from burning gobs of gaseous material in the vicinity of Sol, our Sun, then cool and settle by chance into a variety of orbs, whether comprised mainly of rock, ice or gas. By contrast, the system was planned and materialised into a superbly detailed complex that is managed in force and form – but never by chance.

However we are not saying there is no disharmony, no difficulties to be navigated. Balance is ever attempted and sought after; perfect balance is never attained – true for humans, true for stars and for planets alike. Universal motion is a constant at all levels. The benefit of balance is easy for most of us to grasp intuitively. The problem with *perfect* balance is that if it should happen, motion necessarily ceases and stagnation would be the result. This cannot be permitted so it doesn't occur.

We might at this point apply Semiosis – in other words, ponder on how the patterns we've talked about affect our sense of meaning. If we accept patterning as real, we cannot easily maintain our concept of humanity as we once did. Humans are part of the universal fabric. On the one hand they are definitely not Lost in the Stars yet also, because of their very inception, they can do nothing but obey all universal laws. Humans are not potentially masters of the universe but instead, just one set of uniquely textured pebbles on the great and glorious galactic beach. So once again we come up with the value of humility when considering our lives and situations.

While there doesn't seem very much in the way of any information or traditions pointing to this relationship between humanity and Solar Body, we can mention one indicator that was printed in the 1970s. This was a fairly brief 'Communication from an Angelic Source' published by the Findhorn Foundation, a spiritual community that developed within and near the tiny, windswept former fishing village of Findhorn on the Moray Firth in Northern Scotland.

The Findhorn community was started in 1962 by Peter and Eileen Caddy and their colleague Dorothy Maclean, together with the Caddy's three young sons. The adults had already spent many years living as best they understood it, in accordance with God's Will. Eileen received messages from 'the God within' and she and Peter obeyed them to the letter, even when living conditions were very far from easy. She spent long stretches sitting in meditation, most of the night, in the unheated toilet block of the Findhorn caravan park, so as not to disturb her family crammed into their tiny trailer home. The park was where the three friends and three children moved to, after suddenly losing their employment of managing a local highland hotel called Cluny Hill, at the nearby town of Forres.

Peter started a vegetable garden in the very poor soil around his caravan in order to supplement their subsistence income. It was basically beach sand, with whatever additives like horse manure could be collected at no cost from the vicinity. Not only the soil but the gales and frequent freezing temperatures in this exposed coastal environment seemed to cast the garden as an exercise in futility. But Peter was the epitome of energetic faith and kept working at it as well as encouraging the others.

After a few years the community drew to itself many other interested folk, quite largely due to the amazing fecundity of this garden, where '42-pound cabbages' – which astounded curious visitors and professional gardeners alike, together with subsequent generous BBC TV coverage – became symbols of their success. An iconic image from the garden in those days was the photograph of a rose bush, blooming beautifully in the snow.

The community was happy to explain to anyone who asked for the secret of this prolific growth: They said they had communicated with and enlisted the help of the Devas, which they understood to be the subtle, living forces that underlie and sustain all expressions of vegetative (and other) planetary life.

When I visited the community for a three-month stay in the northern winter of 1977-78, the emphasis was changing from growing plants to 'growing people'. By then the organisation was flourishing with scores of members. It received income from visitors and donations and had even taken ownership of the same Cluny Hill Hotel from which the founders had been dismissed years before.

Spiritual teacher David Spangler was part of this expansion and he lived at Findhorn for a while, offering his own brand of knowledge and understanding, especially regarding the community's significance in the global New Age movement. David received spiritual messages from what he understood as various esoteric sources, some of which were printed and published by the community.

Around the turn of the decade to 1970, there had been an emphasis within the Findhorn community and in some spiritual circles generally about UFO sightings and their significance. Were UFOs from outer space or rather, an inner phenomenon? What did they portend? One of Spangler's messages dealt with this in an interesting way.

The stated angelic source of the message named itself as 'Elouai, a builder' who 'stood with a company of others outside your planet'; whose home was 'within the higher dimensional vibrations of space' and who were 'brought here long ages ago as companions of the god who has created this Solar System'.

The message became more specific: '...there are in the body of the Solar Logos, which is the Solar System and all its planets on all their levels of being, those forces and energies which correspond to waste material'.

Elouai then added that, for various reasons, 'Terra (Earth) was set aside for the special task of being the purifier for your Solar System for a period of time. Left in their exposed state, these energies had the power to impact harmfully upon the sensitive fabric of the solar Being and upon the other planets and their life forms...' And later in this message: 'Your planet became an arena for the interplay of the forces of evolution on many levels'. Then a climax: 'Earth became analogous to a kidney in the body of the Solar System, regulating and transmuting the energies flowing throughout the system'.

It is interesting that Earth is considered the physically densest of the planets, with mean density of 5.6 g/cm^3. This is due to not having gaseous components and all its mass being concentrated and compact.

Spangler's message might have been intimating that Earth's makeup, which could possibly suggest the presence of some lower frequencies relative to the rest of the Solar Body, was suited to transmuting and reintegrating the 'waste' he spoke of. His message had then further explained: 'In the evolutionary movement forward through time, this residue which is left behind must be gathered up and purified and then returned to the creative cycle; it cannot be allowed to accumulate or to express itself within the evolving body as centres of unintegrated and separate energies existing out of timing and out of place'.

Not many people are talking along these lines, but then neither is all the available knowledge spread abroad widely, even in this era of unprecedented information accessibility.

Assuming that Earth is currently patterned on a particular organ, is it the kidney? I don't know and perhaps we don't need to know. The acceptance that there is a specific relationship between organs and planets, can still help us to understand our dependence on the forces that created all of them. Grasping that relationship will help us become more effective citizens of our Solar System, our Body Corporate.

COMING CLEAN

Writers cannot prevent becoming totally exposed to the perceptive reader. What is in the mind, the heart and even the soul will somehow inform the written word. If the writer tries to hide herself or himself, that too will be on display. Best then, not to attempt concealment. It might even be appropriate to come clean about one's motives, intentions and even the driving force that led to one's putting fingertips to keyboard. This makes me think of John Fowles and his best-selling novel, The French Lieutenant's Woman. He fills many pages with thoughts and beliefs about mid-nineteenth century English society, his setting for the story, along with a number of musings and self-confessions. The novel was backed by very fulsome research. My field of play is quite different from his; yet our intentions and methods are not so dissimilar.

John had the great advantage of dealing with the not-too distant historical past – and within the boundaries of his own predilection, the past of his native country to boot. My research tends more towards the as-yet formless future and accepts a very wide stage, a variety of internalised information and a version of prophecy that is deeply felt although of course – most likely less 'respectable'.

We all like to conjure the future to some extent: deciding whether a house in this neighbourhood or the next will be better suited to our needs, whether we want to be an army captain or an architect, whether person A or B would make the better partner. My future presses on me, despite the obvious fact that I cannot know for certain what it is. Nevertheless, that future lives as surely as does my past. Increasing knowledge and acceptance of The Solar Body Corporate is part of it; the livingness grows, becoming as real as the memory of many significant personal events.

In Australia, a body corporate has a statutory responsibility to keep common property in good condition and I feel we must extend our understanding of this Solar System to embrace a similar pattern. It leads to envisaging a time when we view Sol and the planets as an environment modelled somewhat on, shall we say, a high-rise block of housing units, where all the inhabitants benefit from mutual obligations carried out by everyone; where the actions of each resident must impact not only upon themselves, but on all of their neighbours.

This model throws a spanner into the works of our current beliefs and expands our sense of responsibility to previously undreamed of heights, yet it maintains the overall principal of oneness that informs the universe.

We have looked at the way each planet is unique in so many aspects such as size, spin, composition, temperature range, atmosphere, topography, weather systems. Another way of saying this is that each has its own highly complex energy signature.

A mythological future self is being asked to tell here what it knows or more accurately, what it is permitted to know and that depends upon what it is permitted to *be*. To an infinitesimal degree, that self must be a part of the Solar Body Corporate – so may it not speak?

LURE OF THE BODY CORPORATE

We start with a star. One lantern in a galaxy of light. We call it Sol. It is larger than 90 per cent of the other stars in the galaxy, yet would be completely dwarfed by some of the stellar giants. More importantly, Sol is powerful enough to fulfil its purpose. Its super high frequency, supreme intelligence, creative power and total energy output is inconceivable to human minds.

How did this nuclear maelstrom originate? Let's hold a truce between scientific and spiritual/religious opinions; they are both in search of truth as they see it and arguments are unlikely to be of much value. The real truth is that no-one knows what life is, whether defined by us as organic or inorganic in structure, and life's response to that ignorance is to continue, directly and inexorably as before.

Back to the question – how is a star born?

Like everything else, Sol was formed in obedience to a universal Command and the medium for this to occur was primal energy or primal matter – the terms are interchangeable and as we said earlier, we can reference Einstein for that. A vortex was projected which became our Sun – although a better description, already given, for it is a Solar Logos which manifested in physical and non-physical garb. The command to this Logos instructed it to proliferate further. The requirement was for nine organs, or nine planets. The way this happened was with the initial projection of a primary planet, the one we call Jupiter.

It is no random coincidence that Jupiter is the largest of the planets, or that it produces so much more heat internally than it receives from the Sun, or that it has the most moons. (Jupiter of course is the Roman name for the king of the gods so is aptly named.) Jupiter's Planetary Logos was charged with the responsibility of projecting (or manifesting or creating) the other planets and it did so: four to its left – Saturn, Uranus, Neptune and Pluto – and four to the right – Mars, Earth, Venus and Mercury. So we can call Jupiter the primary planet.

While this is all hardly mainstream, Emmanuel Velikovsky in his 1950 publication Worlds in Collision wrote that ancient Greek authors described the birth of the goddess Athene or Athena (Velikovsky equated her with Venus) which 'sprang from the head of Jupiter' (Jupiter being the Roman equivalent of the Greeks' Zeus, also king of the gods).

Velikovsky had some very unusual ideas, to say the least, but to be fair he was a dedicated researcher. A website commenting on his theories was willing to affirm the following: '...the curious stories of the ancient Greeks, that Athena (Venus – eventually, Aphrodite replaced Athena in this role) was born from the head of Zeus (Jupiter) and of the ancient Hindus, that Shukra (Venus) emerged from the mouth of Shiva (Guru or Jupiter)'.

It's also pertinent to mention Jupiter's mysterious Red Spot, a persistent and very visible, high-pressure region in the Jovian atmosphere producing an anticyclonic storm. The Red Spot has diminished in size since the 1930s yet is still around 16,000 kilometers wide; I propose it is definitely associated with Jupiter's primary planet responsibilities.

Each of the nine Solar System planets that we observe, each governed by its Logos – its energy equivalent – is designed to act as one of nine particular, individual organs in the Body Corporate.

Earlier, we said Pluto should not be discounted as a planet. It is one of the Nine and plays an essential role in the Body, each of which represents a particular quality of experience. In fact, this smallest and coldest planet was most likely the location of the first expression of an essential part of the overall spectrum.

The Planetary Logoi, the extremely powerful nuclei

giving rise to these acting organs, these functions, come from and are directed by the Solar Logos.

Dr. Michael Clarage, who gave a presentation at the Dynamic Earth July 2019 conference held at the University of Bath, UK, had written that 'most of the universe is made of matter in the plasma state, where matter is so energetic that positive and negative charges are free to move separately.'

Dr. Clarage said at this conference there were currents, including ring currents around Earth and other planets; 'very dynamic electrical systems; current flowing in and out; and from what I know now it's not simply a current like flowing through a wire into a something and out the other end – current flows in and out of the North Pole; current flows in and out of the South Pole.' He sketched out these currents around planets, looking much like the lines of magnetic force around a bar magnet.

Of the ring currents around Earth (he sketched one to look roughly equivalent to Saturn's rings) he said, 'I think we're up to about four or five now that we know about; it's very complicated stuff. So let's take that starting point as what we know about our local environment. So put it inside the Solar System. Stars also, they do have it and we will eventually measure it; currents going in and out of their poles as well. The numbers are a little tough to nail down but a planet might have ten to the 7th – ten million amps more or less flowing in; a star might have a billion amps flowing in and out of it.

'You start to see the pattern, you draw the pictures; this is a cosmic blueprint if you will, for how electrical structures are formed. Each star is also following this blueprint but then its sub members also follow the blueprint; the planets are receiving their energy from their star; the star's connected to its source of power, the planets are connected to their source of power. So you have to imagine if not direct currents flowing between them, at least some sort of resonance or induction *but the planets only get their power from their sun.'* (my italics).

'If you imagine one of those being Jupiter, when we look at the sky it is just a pretty small dot but the magnetosphere of Jupiter, the body, the real body of Jupiter is huge. It would take up (*he spreads his arms out*) that's how big the body of Jupiter would look if we had eyes to see it.' (The magnetosphere is the region around a planet dominated by that planet's magnetic field.)

Michael Clarage also had something to say on that day which reflects back to the proposition of a non-random universe:

'We understand planetary formation so little currently that we need a different theory of planetary formation for every planet in our Solar System. One of the patterns in electrical systems we see throughout all of nature is membranes, boundaries; plasma naturally forms its own version of a membrane or boundary. So we'll start with a star; and then we ask how does the star fit in its world, the neighbourhoods that it comes from and lives in?

'Before the advent of the Herschel and Planck (*space*) telescopes, we honestly believed that the stars were randomly distributed in the sky. We believed that, right? And then, once you see the filaments, every star we see in the sky is on a filament, there is no randomness to it if you can see the underlying structure behind it.'

(An interstellar filament may be defined as any elongated structure in the cold interstellar medium which has an aspect ratio larger than approx. 5–10 and is significantly overdense with respect to its surroundings.)

The highest level of each Planetary Logos is primarily responsible for playing its part in maintaining the Body Corporate. If we use the Earth as our example, slightly below the Logos is another level of almost as high a frequency, which we might call the Child of the Logos. From the Child, commands came through so that, on a graduated basis, the planet's rocky form emerged. First it was gaseous and later, made solid. These events were inevitable, but not easy; as mentioned earlier, the process included confusion and turmoil. All Earth's physical forms are forms in energy, not separate from but resident in that which is the Child.

This 'Child' also determines Earth's weather, which as we have noted, is becoming not only more unsettled but at times, much more dangerous. If we then accept a connection between malefic weather events and the recent exceptional rate of climate change, which is widely viewed by the scientific community as a result of human activity, we reach an interesting point. We can easily attribute the Child of the Logos in its responsibility for the movement to cycle change as being the driver of changes to the climate. And we see humanity's actions as promoting and accelerating that change – but the increasingly troublesome series of climate events is not in fact dependent upon humanity's conscious choice. More accurately, it shows the working of an overarching requirement.

To refine the subtlety of this particular situation, some people will have the opinion that Man must cease any actions that are believed to intensify climate change; perhaps they will need to voice their opposition to things such as use of fossil fuels that will exacerbate the problem, or even actively to work against such situations in various ways. Other people will take the position that no such action is necessary or even appropriate, perhaps ridiculing those who take the opposing position. But neither of these groups is using their 'free will'. Free will is not a significant factor in humans. We almost entirely act according to requirement.

Some further explanation regarding that last point may be useful. 'Acting on requirement' by no means guarantees we are aware that we are doing so – in contrast we may believe that we have chosen, without interference, exactly what we want to do. But the requirement doesn't care what we feel, at least not as we might in this context define that word 'care'. In most cases a requirement simply integrates with our energy structure in such a way as to give the conscious mind the impression that it *has* chosen. It can take a great deal of learning to recognise this situation; and furthermore such learning depends on the capacity of our soul, our higher self, to make that possible for us.

The mechanics of the above scenario regarding climate change work like this: Our Earth Logos has obeyed the command to proliferate; to project all manner of manifestations that sit within the scope of what is required by higher Authority. And this does not only apply to the huge varieties and numbers of life forms and so-called inanimate forms we see around us. It is more pervasive than that, including highly significant factors such as human *opinions*. The Logos was required to proliferate and we are assured, it did so. Opinions are so very diverse; they take up a great deal of human energy and very few people, if any, hold exactly similar opinions across the entire spectrum of possible subject matter.

Mankind has always been opinionated but the characteristic seems to be strengthening with the increasing popularity of conspiracy theories, which could be described as opinions compounded. Many such theories are spread by various media outlets that in turn benefit from increased audiences and thus advertising revenues. The entire phenomenon is facilitated as never before, by the almost ubiquitous availability of social media. The whole phenomenon seems to be a necessary, culturally destabilising aspect of age changeover.

This is where subtlety really comes to the fore. Yes, we are deficient in 'free will' and yet we are not simply automatons. We can aspire to new perspectives, we can reach, we can open ourselves to higher inspiration. We can refuse to be limited to the most pedestrian thinking and concerns. We can envisage ourselves citizens of the Body Corporate. We can be inspired (in other words, grow 'in the spirit') by our higher self. The old saying has relevance: 'When the student is ready, the Master appears'. The only question then becomes, who desires readiness?

One possible response to our own opinions is to observe them with a degree of detachment, perhaps wondering about how they arrived on our doorstep and in our mental constellation. Are they immovable, set in stone? Are we slaves to them? This process could uncover new perspectives. We are not totally locked in to our current state.

Now let's move on from our own particular, individual microcosms to the larger perspective. Each planet has its own way of participating in the maintenance of the Solar Body, just as each utilises its own particular method of proliferation within its own jurisdiction. The proliferation process is one of stepping down what originated as primal energy – recalling that this is the name we give to the ultimate frequency. For example the process can give rise to multiple, cascading expressions, each level being of lower frequency than the last. This includes, where called for, some of the very lowest and slowest of frequencies, which give rise to physical matter.

How these methods apply within each system is described in terms of number. As far as Earth's own operation and manifestations are concerned, the number is three. 'Three-ness' is common in our religious and spiritual systems; for example, Christianity has Father, Son and Holy Spirit while Hindus have Brahma, Vishnu, and Shiva. Other expressions of trinity can be found across the planet: we are very obviously looking here at a global principle.

It came as a complete surprise to me, to learn that denial of the Holy Trinity was actually illegal in Britain until 1813, showing the depth of belief that's been assigned to this idea. (Presumably the reason for the law was to protect the second member of the Christian Trinity, 'God the Son' – Jesus – as a basic reason for the Christian faith, yet it still demonstrated a powerful investment in the number three.)

Isaac Newton, who had such an enormous part to play on our modern scientific outlook, did not like the Trinity concept and took the risky step of supporting a critique of the doctrine, published in 1712 by one of his followers, Samuel Clarke. Newton was hardly one-dimensional and he not only favoured the dimensions of physics, mathematics and astronomy. He also searched for the elusive formula by which base metals could be turned into gold; he attempted to find the elixir of life such as could keep humans forever young and healthy; and as we have noted earlier, in his own way he was deeply religious.

In the second edition of his *Principia* in 1713 – the first edition appeared in 1687 – among the other alterations and additions, Newton appended a General Scholium (explanatory notes) at the conclusion. This document was deliberately difficult, even obscure. It had the intention of presenting some of his more controversial, even heretical, ideas only to those who had the sophistication and background to unpick them.

For example, he considered the system of planets (which orbit in the same direction on a near plane –the ecliptic) and the comets (which revolve in highly parabolic ellipses in every direction) must have some other explanation than a purely mechanical one. He had become disaffected with such an absolute mechanical philosophy, which he believed had introduced a dangerously secular view of the physical world. Instead, he was eager to reinstate 'active powers', spirit and God to physics.

Newton, who seemed to live in a rarefied atmosphere generally not shared with others, nevertheless believed in One God and a personal god at that. It was among the paradoxes of human history that as a man of such strong spiritual faith, Newton with his ground-breaking scientific and mathematical discoveries made during the 17th and 18th Century 'Age of Reason', would inevitably tend to diminish the power of religion.

Returning to our theme of the Three, perhaps it was because Newton's thinking was on so large a scale that he had skimmed over the 'three-ness' of our planetary environment in his drive to grasp 'the One', or in other words, that principle which he regarded as the ultimate reality.

Earth's trinity principle can be expressed more precisely

by describing each of its interlocking stages: 'Command principle, Linking principle and Applicating principle'. We can describe how these three operate together in practice: In order for a physical manifestation to appear on our planet (it will apply to the non-physical as well) there must first be a command for that to happen. Commands by the way can *only* be made on requirement (otherwise universal chaos would ensue) and then to issue from an entity that acts to generate a result that is lower in frequency than itself.

It has probably become apparent to the reader that 'command' and 'requirement' are virtually identical in meaning. We could consider that a requirement may involve more than a single Command to be complete. And should there be the valid question of 'Required by what or whom?' the simple answer is: By the great Unknowable source that in its very essence *is* this universe.

The Command will generate a nucleified energy event sufficient for the process to occur. The nucleus can then trigger the application of what is required and permitted by that Command. Clearly, these are two of the steps in the trinity process but there has to be a facilitation by an intermediate, third force. The third force is the Linking principle (the Link) which shines directly from what we called the Child of the Logos; it knits together the Command and the Applicating principles.

The linking function, unlike the other two components of the trinity principle, does not diminish in intensity even when the required manifestation is itself of relatively diminished frequency, such as a bee colony or a bacterium – though both of those in their own way are just as wondrous in their design as a galaxy. The Link, in its role of integrating command and application, is the final arbiter of what is permissible as a projection on the planet. It acts as a check to confirm that the projection in each case is the one that was intended and sanctioned.

On the higher level of operation into the Solar Body, the number of Earth is four. It is worth noting that Jungian psychologist Liz Greene wrote in the 1970s: '…every trinity must in the end be integrated by a fourth factor'. Jung's concept of personality included four functions of the ego: feeling, thinking, intuiting and sensing. Also, humans have persistently honoured the quaternity of air, fire, earth and water.

Every planet has its own number that assigns to it co-operative membership of the Body Corporate. This is just a taste of the integration we will be able to discover existing across our Solar home.

UNSEEN INTEGRATION

We suggested earlier the likelihood of homeostasis operating across The Body Corporate. If so, how would that take effect? Each planet has its number or numbers, each has an energy impact on the other planets and here we can re-introduce the subject of astrology:

As many things are being renewed and refashioned in this oncoming planetary cycle, the same is due to happen with Man's understanding of astrology. Let's very briefly review astrology's long history and how it is now practised. Then we'll look at how it is destined to change in line with fresh understanding, which will eventually lead to astrology becoming recognised as a truly valid science.

The roots of astrology are associated with our very early civilisation. We've noted claims that maps of the stars existed long before maps of the Earth and that archaeologists have found cave paintings, mammoth tusks and bones marked with lunar phases. Man has long attempted to cope with life's problems by tracking the night sky and its inhabitants— it has even been suggested that the seven visible 'planets' were our first GPS!

When in 331BC Alexander the Great conquered Babylon/Chaldea, it signalled an advance for Greek civilisation and influence. Ancient Greece eventually made what was considered important advances in astrology, along with developments in medicine, geometry, mathematics and philosophy. The modern names for planets and zodiac signs come from Greek literature.

This was within the era of Plato's student Aristotle (384 – 322BC) who gained eminence from his contributions to logic, criticism, rhetoric, physics, biology, psychology, mathematics, metaphysics, ethics, and politics. Although he was not known as an astrologer, as we noted earlier he nevertheless had something to say about the planets.

In the 12th volume of his work on Metaphysics, Aristotle talks about Kronos (Saturn) Zeus (Jupiter) and Aphrodite (Venus). He comments regarding the tradition that these bodies are gods, while noting that at a later date a myth developed claiming such gods were men, or like animals. 'But if one were to separate the first point from these additions … one must regard this as an inspired utterance…' he wrote.

According to Hamlet's Mill, '…an enormous intellectual achievement is presupposed in this organisation of heaven, in naming the constellations and in tracing the paths of the planets… They are not mere projections of a troubled consciousness: They are attempts to portray the forces which seem to have taken part in shaping of the cosmos… One might also say that while the fixed stars represent the kingly power, silent and unmoving, the planets are the executive power.'

In the second century AD Claudius Ptolemy published *Tetrabiblos* (Four Books) containing core techniques of astrology. Some of his methods are used to this day, including what are called 'aspects' or the angles between planets, sun, moon and other significant parts of the horoscope.

And when the astronomer Copernicus advanced the theory that the Earth travels around the Sun, he dedicated his main work to Pope Paul III (1468 –1549) who was said to arrange his day's work according to astrological principles, which must say a lot about the influence of astrology in those days.

Since then, astrology has had its declines as well as its renewals, including interest from famed psychologist Carl Jung (1875 – 1961) who published his research into statistical analysis of astrology in his book Synchronicity. He accepted the notion of the astrological Age of Aquarius and suggested that this period could 'constellate the problem of the union of opposites. It will then no longer be possible to write off evil as the mere privation of good; its real existence will have to be recognised.'

It has to be said there are several difficulties with the astrological idea that 12 visible constellations, as they each come into eminence through 'precession of the equinoxes', will be the harbinger of a different 'age' or type of consciousness. On the one hand, this is quite popular, as demonstrated by the continuing interest and belief in the New Age of Aquarius. But one reason for scepticism is that there is another Zodiacal constellation to be seen in the night sky that is not included in the hypothesis.

The 12 'astrological' constellations, plus an extra one called Ophiuchus (the serpent bearer), are more or less to be found along the Zodiac band, which extends approximately eight degrees north or south (as measured in celestial latitude) of the ecliptic – the apparent path of the Sun across the celestial sphere over the course of the year.

These constellations, described by their astrological sign, do not align neatly into 12 almost equal periods during the 12 months of the year as generally depicted by today's astrology. The constellations are not only of different sizes in terms of the area of sky each takes up, meaning their coming into prominence should be of varying periods, but if we include Ophiuchus, we should divide the year into 13 periods not 12. Another complication is that, again because of the effect of precession as already described earlier, the constellations are in any case no longer to be seen in the positions where astrology normally says they are. As a result, we have to accept that astrology as currently practised is logically imprecise. It is largely a symbolic system and can be understood and judged only on that criterion.

In recent times astrology has also gained a poor reputation by the practice of various media publishing astrological predictions and advice by zodiac sun-sign alone. (One website claimed that surveys show around 90% of the population – presumably applicable to the USA – knows their zodiac sign, and almost half read their horoscopes regularly). In doing this, they reduce the entire human race at any given time to the rather unlikely position of only 12 possible types of experience.

On and off for thousands of years we have been employing astrological techniques in the hope of discovering what our personal strengths and weaknesses might be, how to relate more effectively with others and to become aware of challenges and benefits on the horizon. But mostly we haven't understood what can be discerned in a broader sphere, a more inclusive picture of global affairs. Could we do so, it might help us take steps to manage them better.

Astrology like everything else is a matter of energy dynamics and we are told that its basis is inherent in the Solar Logos itself. This mighty – we might almost say almighty – Logos is in control of its entire body, The Body Corporate; and it is fully aware of the influences shed by each of the planets.

Through astrology, or any other technique hopefully offering insight or prescience, we nowadays almost exclusively concentrate our attention on our personal, Earthly lives. Mankind will learn that this is only a minor dimension and there are larger perspectives awaiting our concern.

The nature and the corresponding influences of each planetary Logos differ and the frequency spectra involved are enormous. The planets with their individual spectra interact and as a result they can have effects such as cancellation, addition, partial cancellation or partial addition of the frequencies; with the result in each case giving rise to a particular influence. We then have the situation where each of the planets, through their governing logos, exerts an influence across the Solar System. As the influences either oppose or build, they impact on every part of the Body Corporate.

With that larger perspective, we may look forward to astrology becoming a valid science. When we finally uncover a deeper knowledge of it, we may expect much more accurate and worthwhile indications that give insight into conditions and movements that are of truly global significance.

While astrology is the practice of treating the ever-changing patterns of planets as seen from Earth in terms of offering potentially valuable information, we can also say that our incarnate bodies are clearly patterns of our Planetary Logos. Additionally, each Logoic body is a complete pattern of the Solar System – giving rise to and justification for our term of Body Corporate. As part of this close patterning and integration, there is necessary interaction and we are reminded of the human body with its various interacting, complementary systems including hormones and particles (such as nutrients, oxygen via haemoglobin and waste materials) that travel throughout the network in order to maintain health.

Each of the planets must also deliver elements of their own particular energy expression to the others and *no physical means are required.* The process by which these projections occur is called spawning. The spawnings are lives with functions that are directed towards what is required to fine-tune the integration of the Body Corporate. The easiest way to describe their value is as catalytic agents. And due to the diversification involved, there needs to be multiple spawnings. 'Outer space' in our planetary system is neither empty nor inert, which would be obvious if only we could become aware of it in energy terms.

The responsibility for the furthering of our planet's own projections is assumed by a frequency level that is just below what we have now termed the Child of the Logos. We might name this next level – possibly it's a little clumsy yet nevertheless quite self-explanatory – the Child of the Child. Every manifest object, whether so-called animate or inanimate, that we see around us is due to this slightly lower level of energy intensity. Effectively we could call the Child of the Child our planetary god. Another possible label for it is the Demiurge, derived from *dēmiourgós* – 'craftsman' in Greek and introduced as 'the divine artisan' in around 360BC by Plato, once again in his book Timaeus. The Demiurge has been acknowledged by various groups through the centuries. For some it has been considered a subordinate to the Supreme Being and produces all the physical things of the world, including human bodies.

Whatever we call them, these three super high-level intensities, the Planetary Logos itself and those two other functionary nuclei, Child of the Logos and Child of the Child, stepped down to just a little below the Logos' super-elevated frequency range, are way, way beyond our everyday understanding. I think we are quite safe in re-emphasising that.

In passing, we might also say something to help resolve the question of defining 'animate' compared with 'inanimate'; or 'organic' versus 'non-organic' manifestations: it may be easier than we think. We need only to recognise that *nothing can exist without a continuing energy flow*. (Even rocks, seas, tables and chairs, cars and custard.) And we might then accept the standpoint that energy – basically inexplicable but existing everywhere – *is* indeed Life. Then our definitions become conveniences only.

Another way of looking at this affirms each projected state is penetrated and permeated by all its pre-states. That is a consequence of the fact there is no spatial separateness in energy – everything is within everything. Despite appearances, there is a great deal more fundamental coherence than difference.

To reiterate, each Planetary Logos has its distinctive function within The Body Corporate and the projections commanded though any one of these Logoi must be in alignment with the structure and function of that particular Logos.

Should it happen that scientists never discover life as we know it on the other planets and/or their moons, this does not mean that life doesn't exist in these environments. Rather, it means that we don't have the breadth of insight to grasp that life on each planetary body *must* always manifest and operate in accordance with the range of its energy environment; and its projections *must* be different from proliferations that are produced by all the other planetary Logoi.

The situation is hardly surprising if we consider the

required patterning with the organs of the human body: We would not expect the liver and the lungs to be anything other than very different from each other if they are to perform their separate functions correctly. And we might add that the proliferations of life on the planets other than Earth may very well not include *any physical forms*. (The Body Corporate doesn't follow our so-called rules.)

If we really think about it, how *could* there be our definition of physical forms of life on say, Mercury, with the tremendously extreme temperature ranges we have earlier reported as existing on that planet? Then turning the focus around, why should there *not* be non-physical life forms on Mercury? And in fact, there *are*.

This might be the right moment to describe one more element of the Body Corporate. The system is not yet complete, which is to say, there are future planets in gestation. We are told there are sources within the Solar System which 'are observable on radio telescopy' but not recognised: meaning 'the story of this system is far, far from being completed'.

Earlier, we briefly mentioned the ongoing changes that will be made in the human frame. It will eventually become so evolved, so refined in frequency that it will no longer be physically solid. Remembering that the planets reflect the human organs, there will of necessity be changes in both of these energy expressions as part of ongoing evolvement.

And as we look at a human body relating to the Solar Body, we might ask ourselves: if there is a patterning with organs, might there also be an equivalent with glands? The answer is yes, except the glandular equivalents are much less obvious. Throughout The Body Corporate there exist what we might call energy glands, yet unlike the planets they are not visible.

Several religious or esoteric systems, well-known on Earth, teach a correspondence between ductless glands (the endocrine system) and invisible yet potent energy centres, most commonly called chakras (from 'cakra' in Sanskrit) that underlie their operation and which play a vital role in maintaining human life. The energy structure of each planet follows this same pattern. Due to the patterning, there is a direct impaction upon each planet by these energy glands, chakras or force centres. The exact number of them that are applied depends upon the requirement for each particular planet.

Given this information, perhaps the following statement from Dr. Michael Clarage, again speaking at the 2019 Dynamic Earth conference, might be worth considering:

'There's also these other blobs that we see now and as a scientific community we're bounded by what we already know about. And so they have a name; they're called protostars. Anytime somebody uses the word 'proto' - you know they don't know what they're talking about.' (A trawl of the internet suggested a protostar is generally believed to be a star in its early formation. But the hesitation of scientists to stake their reputations on that definition seems not uncommon.)

Dr. Clarage on protostars again: 'So I don't think they're stars – they might be, I don't know. The point is, they might be something else, they don't have to be on this track that we call "be a star"; they could be something else that is needed in the interstellar medium.'

(And to stir the pot, the author of this essay suggests: What about the possibility of some of these so-called protostars or something like them being the equivalent of 'systemic glands', or 'systemic chakras' or even new planets in early formation?)

Within our Body Corporate, the 'glandular' centres are overseen and controlled by a hyper-intense energy source we shall name the Distributor which is physically located within the Sun, although functionally distinct from it.

The Distributor, an energy facilitation currently unknown to our science, is an entry point for forces coming from beyond the Solar System that are absolutely necessary for the system's correct function. For example, *all* Commands appropriate to our system must access it through the Distributor. And unlikely as the following may sound to our ears, without the invisible and to date undetected chakras/force centres galvanising The Body Corporate, without those centres constantly providing energy nutrient to the planets, not a single Planetary Logos could exist. This would mean – no planets; just as without our own force centres to energise us – no humans. Once again, patterning is shown.

At the same time, we shouldn't over-simplify that patterning, because the energies from the force centres are not only delivered separately and individually to each Planetary Logos, but there is a tension between every Logos and its force centres that takes perhaps a very, very long time to resolve. The planetary Logoi have no control over the results of those forces that beam into their projections. Tension occurs due to the force centres projecting ideas, teachings, beliefs; these forces arrive unasked by each planetary Logos and initially there is a rejection.

This explains how it is that despite their initial clarity and effectiveness, many excellent spiritual teachings that have arrived on planet Earth, quite soon become corrupted and fail to reach their promise – which includes their potential for lifting mankind from ignorance, hubris and selfish separateness.

How and why could such a thing happen? We might reply that Man through its immediate projector has to learn from the experience of the dual forces that reside everywhere. This is a complex process involving the soul level that stands behind humanity and the results of the process are not for mankind's benefit alone.

If we name that soul level as the Parent in energy, it might help us to complete a mental picture of the chain of frequencies 'up' from ground level (i.e. frequencies elevated beyond our physical manifestation). The sequence is from Man; the energy Parent which gives existence to (spawns) that physical expression; the Child of the Child; then via the Child of the Logos to the Planetary Logos itself.

Whenever a person (any human anywhere) has an experience of the twin forces, whether ostensibly 'good' or 'bad', the energy result essences back to the person's Parent and eventually from there all the way up the chain. You might have noticed we said the result channels *via* the Child of the Logos to the Logos itself. This is because the human event will have its effect, however individually miniscule, on the other levels; but never on the Child of the Logos itself. The Child is the Link and the Link is never altered by anything other than itself. This is actually self-evident because the Link represents the single, great Universal Command – if anything else was able to change it, that definition and that function could not apply.

One useful way to better understand the Child is to see the similarity between its operation and that of the human autonomic nervous system. As a reminder of that system's meaning and function: it is a component of the peripheral nervous system that regulates so-called involuntary processes including heart rate, blood pressure, breathing, digestion, and sexual arousal. It contains three anatomically distinct divisions: sympathetic, parasympathetic, and enteric – otherwise, briefly, 'freeze, fight and flight', 'rest and digest' and the actual management of digestion. In other words, the Child can be seen as a regulating principle for Earth, as long as we recognise that the nature of such regulation only happens by and through Command.

We can make one or two more observations. If you consider a particular day in the life of a lone human being and make a guess at how many of the above-mentioned happy or unhappy energy events might occur; and then multiply that figure by the experiences of the other eight billion humans, you get some idea of the complexity of the entire procedure. And this is without taking into account what will occur as the energy result proceeds up the chain, with its processes being dealt with and affecting each level in turn.

In passing, it's very likely that those higher levels will gain more benefit than we ourselves do, thus providing us with some comfort that our existences can be seen as a greater service than we may have deemed possible. The interplay of forces in people will eventually essence back to higher and higher frequency degrees, thus assisting the continuing requirement for universal difference – which is the basis for universal evolvement.

Until the Logos becomes sufficiently aware, which can only occur when universal laws permit, the previously mentioned esoteric teachings arriving via the planetary force centres will languish and not become established as hoped for and expected. This principle even applies to the words you are reading now, so you may reasonably ask - why should I the author, persevere with this project?

Because there is a change occurring. Despite massively increasing global turbulence, there is something different in the air that speaks to renewal; the gossamer impression of an upcoming creative adjustment. We might well expect a revolution in knowledge, in belief and in understanding. If we harbour a poetic inclination, we could describe the change as a New Heaven and a New Earth. It is called the Sixth Cycle.

GOING BEYOND

Unsurprisingly, as this author is an Earthman his first concern must be other Earthmen and women. At the same time, he is only too aware that Earth and the Solar Body itself is a very, very small part of what is most likely an infinitely large universe.

The human mind is not well geared to process the idea of infinity. We might use the word easily enough but can we encompass and apply its meaning? This is where the well-used term 'the mind boggles' comes into its own. Infinity hardly registers beyond a label for the purely theoretical. It sits there in a kind of rude ignorance that defies any useful application. But like many other nuisances it won't entirely leave us alone.

If we should simply ascribe finiteness to the universe, that doesn't solve the problem. That which is finite has boundaries and so what happens beyond them? Infinite finiteness?

Yet perhaps infinity is not so difficult. Perhaps, being as we are miniscule expressions of Earth's Planetary Logos, we already carry with us the infinite nature of life. Our physical outer garb will die and return to the elements of its construction, but through our soul – our Parent in energy – we will continue in other guises for a length of time that may well be timeless.

Within the universal context we are investigating, we have to admit we are only taking a place in the galactic kindergarten. The Body Corporate we're looking at is a very small corner of the greater universe. Our solar environment is intricate and fascinating, replete with complexities and speculations that might take many future generations to unravel. There are levels of knowledge we may not ever be able to uncover.

While we find it virtually impossible to conceive of the power, abilities and stature of any of the planetary Logoi, how much more demanding it would be to grasp the extent of these qualities in the Solar Logos. Something that might be a little easier for us to contemplate is the relative relationship between the planets and Sol – in fact it could be said that the Solar Logos *is* this System, this Body Corporate, this entire entity.

Sol's age has been calculated at 4.603 billion years and the star is anticipated to continue for another five billion years or so. Then the expectation is that it will become a red giant. If so, according to science, the core will shrink but the outer layers of the Sun should expand enormously, until its size reaches the orbit of Mars. What the entire system will look like by then can only be conjectured, but if we accept that nothing happens by chance, everything being part of a universal plan, this could be the Body's complete demise – function fulfilled.

We might ask if there is anything above the Solar Logos in power and authority. The answer is yes; we could take it as a principle that everything at every level of existence is enfolded into a greater and greater reality and it only stops when we reach the Unknowable.

We can say our Body Corporate is just one system among a group of nine Solar Systems that fall under the responsibility of the same controlling parent, which is the binary star Sirius. It is comprised of Sirius A and Sirius B.

We could begin by looking further at this star that via our Solar Logos ultimately sustains not only us humans but all the degrees and the denizens of our Body Corporate. We cannot identify the other eight systems but we can assume they will each operate and cause projections according to 'local' principles that are different from ours. Most likely for quite some time at least, these other systems under the jurisdiction of Sirius will be 'none of our business'; in any case, learning about our own is going to take up no shortage of our attention.

We might guess that there are similarities (patterning)

in the relationships between those other systems and our Body Corporate and Sirius. But to help inform that speculation, we can only present the bare bones of what is known about the linkage between the brightest star in the sky and our Sun and planetary system.

Sirius is the alpha star – first or brightest star – in the constellation Canis Major ('Big Dog', thus Sirius is well-known as the Dog Star). Its name is derived from the Greek word *Seirios*, meaning 'glowing' or 'scorching'.

Sirius has had particular significance for several cultures over the centuries. The heliacal rising (i.e. rising above the horizon just before sunrise) of Sirius marked the flooding of the Nile in Ancient Egypt and the 'dog days' of summer for the ancient Greeks (hot sultry days were associated with bad luck) while to the Polynesians, mostly in the Southern Hemisphere, the star marked winter and was an important reference for their navigation around the Pacific Ocean.

Sirius A is twice as big as Sol, having 2.06 solar masses and around 1.711 solar radii. It is 25.4 times more luminous than our Sun. While the temperature on Sirius A has been recorded at 9,667 Celsius, the dense white dwarf Sirius B was found to be much hotter at 25,000 C.

Although so much smaller than its larger companion (a little less in size than that of Earth) the white dwarf star Sirius B has a mass that is enormous, only slightly less than that of our Sun. This was determined using one of the key predictions of Einstein's theory of General Relativity. As a result, Sirius B's powerful gravitational field is no less than 350,000 times greater than Earth's, meaning that a 68-kilogram person would weigh 25 million kilograms standing on its surface (although of course they would no longer be standing – or even recognisable as a human of any weight).

Sirius B is invisible to the naked eye. This is due to its small size, faint luminosity (about 10,000 times less than Sirius A) and the closeness to its larger, super-bright companion. One astronomical website pointed out: 'You won't be able to differentiate the two unless you have a really big telescope under exceptionally dark, clear skies'.

It is therefore intriguing that the so-called primitive inhabitants of villages in Mali, the former French colony in sub-Saharan West Africa, were found to have identified several features of the Sirius system, especially Sirius B, when investigated by two anthropologists over many years in the 1930s, 40s and 50s. It is a little-known story even now, despite publication of the book The Sirius Mystery by Robert K. G. Temple in 1976.

Marcel Griaule and Germaine Dieterlen carried out most of their research with the Dogon people of Bandiagara, situated not too far from the legendary city of

Timbuktu. (Three other Sudanese tribes, the Bambara, the Bozo and the Minianka as well as some others, were also involved.) The French scientists were clearly diligent and sympathetic to the local culture, as shown by the fact that around 250,000 tribespeople massed for Griaule's funeral when he died in 1956.

The anthropologists came up with some remarkable material during their extended time in conversations with especially, Hogons – the title given to the Dogon's spiritual elders. The Hogon Ongnonlou had described the Sirius system as part of a 'slight acquaintance' with an 'extremely secret body of knowledge'. The Sirius system was associated with 'renovating people and by extension, the world'. And according to Griaule and Dieterlen 'the system described here represents one phase of the revelations permitted to initiates'.

While the Sirian system, *sigi tolo*, was so important to the tribe's spiritual life, the truly vital element of their ceremonies was the white dwarf Sirius B, *po tolo*, which for convenience the French scientists named Digitaria. Temple gives the reason for them using this label: he explained that '*po*' is a cereal grain whose 'official botanical name is *Digitaria exilis*... what is significant about the *po* grain is that it is the smallest grain known to the Dogon... this represents the tiny star'.

When Griaule and Dieterlen recorded their research,

they described Sirius B as 'hardly visible'. They were wrong. As mentioned, Sirius B definitely cannot be seen by the naked eye and the Dogon did not have telescopes. According to Robert Temple, the Dogon knew the star was invisible, but they were equally certain it was there nonetheless.

The Dogon priests also believed there was a third star in the Sirian system (as well as a possible planet). Interestingly, a Wikipedia entry says: 'Since 1894, some apparent orbital irregularities in the Sirius system have been observed, suggesting a third very small companion star but this has never been confirmed'.

Here are some other things the Dogon should not have known, but did: Digitaria's orbit is an ellipse, not a circle around Sirius A; moreover the ellipse is perpendicular to the horizon and when Digitaria is close to Sirius A the latter appears brighter. Conversely, at its most distant from the bright star, Digitaria gives off a twinkling effect (again noting the proviso – *none of this* is visible with the naked eye). Temple commented that whereas the knowledge of planetary orbits being elliptical has been recognised by scientists for hundreds of years, it was hardly likely that this would be known to the average European, let alone members of the Dogon tribe.

Temple further pointed out that the Dogon knew the Earth turns on its own axis and that they also 'knew perfectly well that it is the turning of the Earth on its axis which makes the sky seem to turn around – the apparent movement of the stars from east to west.'

Other Dogon beliefs included: the moon is 'dry and dead like dry dead blood'; and 'Jupiter follows Venus by turning slowly round the sun'.

Attempting not to over-emphasise the weirdness of all this, how did the invisible Sirius B become so important to these so-called 'unsophisticated' tribespeople, who insisted that the star's orbit was situated 'at the centre of the world'? And that: 'without its movement no other star could hold its course'?

The Dogon believed the period of the orbit of Digitaria is 50 years – perfectly correct, although wanting to be absolutely precise, our science usually adds a small fraction of a year onto that number.

Also according to the Dogon, Digitaria rotates upon its axis over the period of one year – I could not find confirmation of this aspect of Sirius B from any internet search. Then I read in Robert Temple's book that he had the same problem in confirming or denying the Dogon's claim and asked Oxford astronomer G. Wegner about it. The answer, when given sometime around the early 1970s, was that modern astronomy didn't know. But on further discussion with Temple, Wegner said although we had no way of determining it, 'A year could be right'.

The Dogon knew that Digitaria was the smallest kind of star, yet they considered it the heaviest. They believed it was built from a metal they called *sagala*, something not to be found on Earth and so heavy 'that all earthly beings cannot lift it'. This would certainly be true of what our science says about the mass of Sirius B.

Dogon teaching was unequivocal: 'God created Digitaria before any other star' and as it developed, it gave birth to everything that exists, visible or invisible.

When we speak of Sirius from an esoteric point of view, a universal energy point of view – a universal management point of view – Sirius B is indeed of a much higher frequency, consequently of greater awareness, of greater power. Just as the Dogon regarded Sirius B as of ultimately the primary importance, so we can confirm that this white dwarf *is* the 'senior partner' of the entire Sirian system.

Sirius B is called The Old One. This is the seat of authority, covering the nine Solar Systems including ours and hence the Earth. That authority is known as the Seven Principles and we might note the parallel with seven ages described for our planet, as well as the enduring significance people have placed on the number seven. Sirius A is known as The Bright One or Birthed One and this star actually manifests those Seven Principles.

How could the Dogon have come up with beliefs that

even somewhat paralleled this understanding? Possibly our biggest difficulty with their culture is in accepting that, despite first appearances, it could have been so highly developed. Germaine Dieterlen wrote: 'The Africans with whom we have worked in the region of the Upper Niger have systems of signs which run into thousands, their own systems of astronomy and calendrical measurements, methods of calculation and extensive anatomical and physical knowledge, as well as a systematic pharmacopoeia.'

Before leaving the fascinating and enlightening conclusions of Griaule and Dieterlen, we should mention that they discovered that the Dogon also knew a great deal about the Solar System and beyond. In addition to their original article, *A Sudanese Sirius System* (translated from the French) much additional information was chronicled by the anthropologists and eventually published in a French language book called *Le Renard Pale* (The Pale Fox) in 1965.

The belief systems of the Dogon as presented in The Pale Fox and other research material, confirm Dieterlen's assertions that they are enormously complex; an English edition of the book included 561 pages of text, photos and diagrams. Thanks to Robert Temple's thoroughness, he included a good deal of pertinent additional knowledge in his book, which gives readers more to ponder on. It included Dogon drawings of Jupiter with four of its moons and Saturn with a prominent ring – once again, all invisible to the naked eye. There was also this statement: 'For the Dogon an infinite number of stars and spiralling worlds exist'.

According to Temple, the French anthropologists as scientists of record, did not attempt any hypotheses to account for their discoveries. All my research into their work confirms this.

The Dogon story didn't sit well with everyone, as you might expect. It seems that in 1991, anthropologist Walter van Beek commented about the Dogon, 'Though they do speak about *sigu tolo* (sic) … they disagree completely with each other as to which star is meant'. Further, van Beek claimed that 'All agree, however, that they learned about the star from Griaule'.

One thing that is not arguable about Griaule and Dieterlen was their amazing dedication in gaining the confidence of the elders, who kept their esoteric knowledge very close and only revealed its innermost secrets to the French couple over their many years-long relationship with the Hogons. These included secrets that other members of the tribes were mostly not privy to.

The 'spoiler' claim by van Beek was published in Wikipedia, but it omitted to mention that Giraule and Dieterlen had begun their field research of the Dogon, which stretched over more than two decades, beginning 60 years before van Beek's comments.

Considering that three of the four Dogon spiritual elders, who according to Temple's book provided the information, were described as being elderly and the other was apparently at least middle aged within the time of the French anthropologists' research, they would surely have died since; consequently they were in no position to disagree with van Beek's claim. Perhaps even more significant is the fact that West Africa including Mali had undergone enormous cultural and political changes; particularly noting Mali's upheavals in governance from 1960 to 1991 – and also because that latter year was a time of unrest and included periodic armed clashes.

Village life in Mali was surely very different by 1991. There is no guarantee that knowledge of the Sirian connection, even if extant and accurately handed down through the generations, was any longer of the same spiritual significance to the Dogon, or indeed whether their entire spiritual life was as important as it clearly had been in the 1930s, 40s and 50s. We owe a debt of gratitude to Griaule and Dieterlein for the careful work in documenting this extraordinary and quite likely fragile local belief system that they undertook so long ago. The highly complex spirituality of the Dogon and the painstaking recording and goodwill of the researchers opens a world of wonder that offers a break from the sometimes overly self-satisfied ideas of our own culture.

I do seem attracted to books that offer extraordinary amounts of research – requiring a great deal of patience to work through. The Sirian Mystery is one of the best (or worst) examples. A challenge is to extract the most significant, succinct information without oversimplifying the details, the arguments and the conclusions. For example, Robert Temple carried out and presented breath-taking amounts of study in the chapter of his book titled Origins of the Dogon (21 pages of small print, including various maps and diagrams).

I chose the following extracts as a reasonably useful summing-up of his account:

'So it is quite clear by now that peoples intimately connected with the Sirius tradition came from Greece to Libya and thence south to the Libyan oases of the Sahara, thence further south-west past the Sahara to Timbuktu and the region of the Dogon, where they mingled with Negroes of the Dogon region and took their local language for themselves, eventually becoming indistinguishable from the local African population in appearance and speech, but retaining their old traditions as their most secret doctrines.

'It almost seems too amazing to be true that we should have begun this book by considering a strange African tribe, then considered similar Sirian traditions in the Mediterranean stemming from ancient Egypt and then be led back again to the African tribe whom we discover to be directly descended from the Mediterranean peoples privy to the Sirian complex.'

The reference to ancient Egypt is hardly a surprise, considering the tremendous importance placed on Sirius by Egyptians of that era. Graham Hancock in his book Fingerprints of the Gods quotes a passage addressed to the god Osiris in the Pyramid Texts – said to be the oldest coherent body of writing in the world – hieroglyphics that are inscribed on the dingy walls of the tomb chamber beneath the crumbling Unas pyramid at Saqqara.

'Thy sister Isis cometh unto thee rejoicing in her love for thee… she becometh great with child like the star Sept. Horus-Sept cometh forth from thee in the form of Horus, dweller in Sept.'

Hancock was intrigued by '…the clear implication that Sirius was to be regarded as a *dual entity* in some way comparable to a woman "great with child". Moreover, after the birth (or coming forth) of that child, the text makes a special point of reminding us that Horus remained a "dweller in Sept", presumably suggesting that he stayed close to his mother.'

It was not too large a stretch for Hancock to suggest this could mean that Sirius was recognised in those days as a binary star, despite him acknowledging what we have already emphasised – that Sirius B could not have been seen without a telescope. (For the sake of accuracy I should point out that while Hancock applies *Sept* as a name for Sirius, I could only find one other internet reference suggesting that name, along with Sothis or Sepdet. But this is probably because we are told that the exact pronunciation of ancient Egyptian is uncertain, as vowels were not recorded until a very late period.)

Sirius has long been associated with the goddesses Isis and Sopdet – the Greek Sothis – and also the god Anubis which has the shape of a jackal, suggesting a fairly obvious connection with Sirius' popular name of 'dog star'. Curiously, this link has been made by several different cultures both ancient and modern across the globe which have also used the label of 'dog' for Sirius, even if they did not know that our name for it came from the Latin word for dog – 'canis' – which title the Romans applied to Sirius' constellation. And as we know, Canis Major is still its official name.

Let's give Graham Hancock a little more of our time. He wrote: 'Temple argued cogently that this high quality technical information had been passed down to the Dogon from the Ancient Egyptians through a process of cultural diffusion, and that it was to the Ancient Egyptians that we should look for an answer to the Sirius mystery.'

Then he added: 'Like Temple, I had begun to suspect that the more advanced and sophisticated elements of Egyptian science made sense only if they were understood as parts of an inheritance. Unlike Temple, I saw no urgent reason to attribute that inheritance to extra-terrestrials.' This reference to possible ETs came from the Dogon belief in the Nommos – amphibious creatures described as coming from Sirius to set up society on Earth; and regarded by Temple to be more or less equivalent with the Sumerian and Babylonian tradition of the fish-men called Oannes.

In fact, Temple's central theme in writing The Sirius Mystery was his speculation about the possible physical travel to Earth of intelligent beings from a planet in the Sirian system. I readily admit to being highly dubious about the likelihood of physical travel by ETs to Earth.

Nevertheless I had a far more positive response to this remark from Temple: 'Nommo is the collective name for the great culture-hero and founder of civilisation who came from the Sirius system to set up society on the Earth'. This is particularly interesting because it accords rather closely with the information I have that spawnings from other planets (remembering our definition of spawnings as being able to travel between planets without using physical means) greatly assisted the social development of early mankind; and which even now may be recalled in myth. We are here relating this process to planet Earth as an organ, which along with all the other planets needs spawnings which act as the equivalent of hormonal infusions to promote healthy operation.

Taking his own direction, Graham Hancock wrote: 'To my mind the anomalous star knowledge the Heliopolitan priests had apparently possessed was more plausibly explained as the legacy of a lost human civilization which, against the current of history, had achieved a high level of technological advancement in remote antiquity.' (Heliopolis was one of the oldest cities of ancient Egypt, close to the Giza plateau and now part of modern Cairo.)

Hancock's ideas are not a new speculation. It is only left for us to query how that 'lost human civilisation' might have sprung up as he believes. No doubt anyone with an inquisitive and discerning mind will sooner or later be able to uncover facts about ancient Egypt that are (1) unexplained and (2) stimulating of curiosity if not awe. Sirius will definitely be in that mix. When we say 'ancient', according to the scanty, incomplete but insistent records, the Egyptian civilisation goes back tens of thousands of years. And when we say 'unexplained', it requires dismissal of the dull work of many mainstream Egyptologists. But why do I claim dullness and how does it occur?

There is a certain arrogance in the belief that human

development has always gone, and only can go, *forwards* in time. This belief consequently abhors any indication that one or more highly accomplished civilisations not only existed in prehistory, but may well have surpassed our own in some aspects. To support this idea, we can again point to the effect of cross-planetary spawnings which can introduce completely new and unexpected elements to a particular planet, region and culture. Should such a scenario be evidenced and substantiated regarding Earth, it means our 'primitive' ancestors were not all as lowly as we generally (and often preferentially) suppose.

There is a type of scholar who would find an outcome like this unacceptably humiliating – so they might wish to guard against it in advance by laying a negative groundwork. By contrast, those of more open mind might prefer to consider it exciting. (We suggest anyone wishing to dig deeper into the tantalising mysteries of ancient Egypt could do worse than read up the carefully researched works of the more adventurous authors such as Hancock, Bauval, Jane B. Sellers and the late John Anthony West.)

Meanwhile there is still plenty more evidence that the binary star Sirius and planet Earth, have strong and enduring connections. For example:

In Hindu mythology, Sirius is the celestial abode of the divine teacher, Guru Brihaspati. This revered sage is associated with wisdom, knowledge, and spiritual guidance. The star is also known as Lubdhaka in Hindu scriptures, symbolising the pursuit of higher learning and the illumination of the soul.

Going from the orient to the occident, a book called The Cosmic Doctrine, written by the historically well-known occultist Dion Fortune (Violet Mary Firth, 1890 – 1946) and published posthumously in 1949, made the claim that: 'this volume of teaching was received from the Inner Planes during 1923 and 1924'. I was hardly surprised to read on and find yet another source that had something to say about Sirius.

The Cosmic Doctrine stated: 'The star Sirius has always been held in esoteric teaching to have much influence on Solar evolution …stars in general and Sirius in particular (because of its special influence) can be considered in relation to Solar evolution – more especially those which, like Sirius, have had a marked effect on it.'

And harking back to an earlier theme, in her book The Training and Work of an Initiate (1930) Fortune suggested: 'We might compare the different evolutions occupying our Solar System to the different programmes coming through the ether on the wireless'. In this she was supporting the idea that wavelength, or its alter-ego frequency, is the sole differentiating factor. (We should note the scientific definition that the shorter the wavelength, the higher the frequency.)

Sirius gets plenty of mentions in that complex tome, Hamlet's Mill. For example, in Mesopotamia and China (and supposedly depicted in the famous Zodiac design discovered at the Hathor Temple of Dendera, upper Egypt, as well) there was a constellation having a 'bow' shape, comprising stars found in our current constellations of Argo and Canis Major. It showed Sirius as the tip of the arrow, or else the target of it. In typical style, Hamlet's Mill comments that 'Sirius-the-target has made more mythical "noise" than any other star'.

The brief accounts given here suggest that knowledge of our intimate connection with Sirius has followed us throughout recorded history and the more specific information that has recently trickled down to us is a precursor to a new understanding of how deep that connection goes – even to the extent that Sirius was ultimately responsible for the creation of our planet and is now monitoring and assisting its progress.

That direct source of information tells us the Sirian forces have observed our planet (and no doubt the others in our Body Corporate) from the very beginning of its existence. This may be a difficult, even an unwelcome idea. But if we are willing to accept a re-framing of the concept, it gives confirmation that we are not alone, not unaided and definitely not Lost in the Stars.

As a planet within the larger Sirian system, Earth is

bound to have the attention of the brightest star in our sky. This is as it should be – it is the function of Sirius to attend to the necessary, higher level requirements of all those suns and planets that are under its jurisdiction.

Those of us who, especially as children, could have been scared by stories and movies of aliens from 'outer space' might still be nervous at the prospect of recognising or even meeting up with our energy elders who hail from a distant star. But these worries are associated with our cramped concepts, our separatist tendencies and our unwillingness to change and grow. Should they ever wear a form, those elders are still, like us, part of the One Life. And while the Sirian forces inevitably are greater, more powerful than most of us could ever imagine, they do not harbour the petty personalities of our race.

There are surely dangerous days ahead for Earth but along with the inescapable insecurity and disruption, including the likely demise of many of our most cherished beliefs about where true value lies, will come a new dawn for mankind.

The Body Corporate eventually has great knowledge, uplift and wisdom to offer us – if we have the courage, open-mindedness and dedication to listen, hear and respond to its ever-developing call.

-ooOoo-

Made in the USA
Monee, IL
04 November 2026

67386439R10069